Books GA
£3.50
25/04/2025

Milk and milk products
From Medieval to Modern Times

Proceedings of the Ninth International Conference on Ethnological Food Research, Ireland, 1992

Edited by PATRICIA LYSAGHT

Published by
CANONGATE ACADEMIC
in association with
the Department of Irish Folklore,
University College Dublin
and
the European Ethnological Research Centre,
Edinburgh

First published 1994 by Canongate Press, 14 Frederick
Street, Edinburgh, Scotland

© The Editor and Authors severally, 1994

Financial help towards the cost of publication is gratefully
acknowledged to The Russell Trust and the Scotland
Inheritance Fund.

Typeset by The Electronic Book Factory Ltd, Fife, Scotland
Printed in Great Britain by Bookcraft Ltd,
Midsomer Norton, Avon

All rights reserved. No part of this publication may be
reproduced in any form or by any means without the prior
permission of the publisher.

Foreword

The great importance of food in folk life and folk life research is recognized in Ireland through the fact that Food and Drink, Meals, Household Vessels and Utensils, all form major headings under Livelihood and Household Support, one of the fourteen main categories into which Seán Ó Súilleabháin's *A Handbook of Irish Folklore* (1942) is divided. Since this book has for half a century formed the basis for both the collecting and the indexing of Irish folk traditions, it is natural that the holdings related to these subjects now in the Department of Irish Folklore (formerly in the Irish Folklore Commission) are rich and on the whole easily available. The entries in Caoimhín Ó Danachair's *A Bibliography of Irish Ethnology and Folk Tradition* (1978) and the supplement to it in *Béaloideas* 48–49 (1980–81), also bear witness to the fact that they have been put to good use by a number of Irish scholars. It is obvious, though, that the Irish material can be but imperfectly understood without a wider international perspective; neither will its full significance be apparent until it is used in comparative research. The very fact that the *Handbook* is modelled on the indexing system used in Sweden in the *Landsmåls och folkminnesarkivet* in Uppsala, facilitates such comparisons.

Against this background, it is natural that the Irish Folklore Commission, and the Department of Irish Folklore, should welcome the initiative which led to the establishment in 1970 of the International Ethnological Food Research Congresses. These have since been held every two or three years, and Ireland has been represented almost from the very outset by Dr. Caoimhín Ó Danachair, and since 1983 by Dr. Patricia Lysaght, who organized the 1992 congress in Ireland.

On behalf of the Department of Irish Folklore, University College Dublin, I bestowed upon the Congress the traditional hundred thousand welcomes to Ireland, and I now heartily welcome the publication of the Congress proceedings.

Bo Almqvist

Contents

Introduction

Part I National and Regional Surveys: Everyday, Festive and Crisis Diet

1 The Consumption of Dairy Products in The Netherlands in the 15th and 16th Centuries   3
   Johanna Marie van Winter (The Netherlands)

2 Milky Ways on Milk-Days : The Use of Milk and Milk Products in Hungarian Foodways   14
   Eszter Kisbán (Hungary)

3 Milk in the Rural Culture of Contemporary Assyrians in the Middle East   27
   Michael Abdalla (Poland)

4 Milk Products in the Everyday Diet of Scotland   41
   Alexander Fenton (Scotland)

5 Milk Dishes and World War 1 : The Revival of Traditional Recipes in Germany?   48
   Barbara Krug-Richter (Germany)

Part II The Urban Retail Milk Trade

6 The Supply and Sale of Milk in 19th-Century Edinburgh   64
   Una A. Robertson (Scotland)

7 The Retail Milk Trade in Transition : A Case-Study of Munich 1840–1914   71
   Uwe Spiekermann (Germany)

Part III Traditional Milk Products and Preparations

8 Cheese-making on the Åland Islands   97
   Nils Storå (Finland)

9 Curds and Pressed Cottage Cheese in Latvia   107
   Linda Dumpe (Latvia)

10 Variations on *Gomme*   114
   Karin Kvideland (Norway)

11 The Use of Whey in Icelandic Households   123
   Hallgerður Gísladóttir (Iceland)

12 Koumiss in Mongol Culture: Past and Present   130
   Joyce S. Toomre (USA)

13 A Swedish Beer Milk Shake   140
   Nils-Arvid Bringéus (Sweden)

14 Porridge Consumption in The Netherlands: Changes in Function and Significance   151
   Jozien Jobse-van Putten (The Netherlands)

Part IV Ice-Cream

15 Ice-Cream in Europe and North America   165
   Laurie Jo Shapiro (USA)

16 'Good Humors' and the 'Good Humor Man'   176
   Molly G. Schuchat (USA)

Part V Milk and Folk Belief

17 Milk and Folk Belief: With Examples from Sweden   191
   Anders Salomonsson (Sweden)

18 Milk and Milk Products in a Woman's World   198
   Ann Helene Bolstad Skjelbred (Norway)

19 Women, Milk and Magic at the Boundary Festival of May   208
   Patricia Lysaght (Ireland)

Introduction

Perhaps it is their very familiarity that causes us to take basic foods for granted. They are so subtly integrated into our normal food structures and meal systems, and so in tune with our culinary world view, that we are wont to underestimate their role in everyday life. We are apt to lift them from our doorsteps, or drop them into our shopping baskets in the supermarket, with little thought for their life history or their role in the process of civilisation.

In the affluent western society modern technology and sophisticated marketing forces are successfully combining to update and upgrade existing dietary staples, and to internationalise our taste in basic foods. Attractively packaged pints of Irish milk—full cream, low fat, super low fat (fortified with added vitamins A and B, and also calcium), skimmed or buttermilk—may now combine with French butter, Danish yoghurt, Dutch cheese, and perhaps German ham, to grace the Irish suburban breakfast table, for example. Thus the basic foods of many countries—especially those in the EC—have become international, and have been adapted and modified to suit the requirements of interregional trade and tastes. Consumers are now more prepared than ever before to eat other communities' or regions' versions of their basic foods, or even to adopt unparalleled dietary staples of other countries into their meal systems.

Conversely, one of the consequences of the internationalisation of food has been a refocusing of interest on erstwhile regional or local basic foods and dishes, the production and marketing of which technological advances and modernisation had forced into virtual oblivion. Some of these have now regained prestige, and have re-emerged in revitalised forms as 'traditional' or 'health' foods both for the home market and the tourist industry.

The basic food profiled in this book is milk derived from milch cows, though milk-based products, such as cheese made from sheep's milk, for example, and a drink called *koumiss*, derived from fermented mare's milk, are also featured. Our central enquiry relates to the role of milk in the life of man—in the food structures and meal systems of diverse cultures in many parts of the world where it is a basic food.

Milk is a vital human food. In a mixed diet it is particularly valuable for its content of high quality protein and easily assimilated calcium, and as a rich source of riboflavin. Yet, it is a uniquely vulnerable commodity and difficult to maintain in peak condition. In some cultures fresh milk was considered a luxury food, while in others it was thought to be unwholesome and unhealthy. In fact it seems that the drinking of fresh milk was the least important aspect of milk utilisation overall, and, consequently, as the wealth of fascinating detail in this book shows, conservation-inspired milk products helped to spread the availability of milk over a longer period of time, giving rise to a range of regional foods and dishes which were integrated into the daily, seasonal, festive and penitential round of rural and urban life.

Indicative of the ancient lineage and the fundamental nature of the wide variety of milk-based products discussed in this book, and the sophistication of pre-industrial conservation techniques, is the wealth of terminology coined to identify every stage of the conservation process, the utensils used, and the resulting by-products, both liquids and solids. The evidence of archaeology, ancient, medieval and modern literature, and the utilisation of milk products such as butter, cheese, skimmed milk, curds and whey in a variety of ways, over wide geographical areas, and in diverse cultures, is testimony to the importance of these products from pre-historic times to the present. This applies also to the complexes of belief, custom and ritual which surround and guide their production, promotion and use.

It is evident from the contributions to this book that the levels and modes of utilisation of milk and milk products varied (and vary) somewhat within different societies and cultures. In many pre-industrial societies the production of milk products was mainly for domestic consumption, while in others, the primary milk products—butter and cheese—had other economic uses as well, notably as a source of barter or trade, or in the payment of rent by small tenant farmers. In some regions, also, these products became items of interregional trade and competition. The adoption of the Swedish-invented continuous centrifugal separator, and the growth of the co-operative creamery movement—in Ireland as well as in other parts of Europe—in the last decades of the nineteenth century, contributed significantly to the development of international trade in milk-based products, but also marked the beginning of the end of the pre-industrial production of milk products, and the way of life which that system entailed.

The challenges posed, and the opportunities offered, by growing urbanization and industrialization in the course of the nineteenth century, gave new impetus and direction to the production of milk and milk products, at both local and national levels. The demand for these products increased dramatically and confirmed their status as cash commodities. The organisation of production to meet demand, and the regulating of the retail urban milk trade, and general milk quality, were some of the major problems facing the emerging milk industry in the late nineteenth century and the early decades of the twentieth century.

Another aspect of milk production and processing dealt with by the contributors to this volume is the role of women in this activity. Historically, and in modern times, dairying has been mainly women's work. Women made an important contribution to the household economy, either in terms of actual milk products for immediate or long-term home consumption, or in cash benefits derived from the sale of milk or dairy products. The economic importance of milk and milk products in the household and farm economy, and the significance of women's role in their production in pre-industrial times, and during the early phases of the mechanisation and the modern commercialization of milk production, is also highlighted in the folklore surrounding milk production and dairying. Women are frequently portrayed both as guardians and 'thieves' of milk 'profit'—the butter fat or cream, and thus ultimately the butter itself—especially during the summer milking season. In either role they are credited with supernatural knowledge, and the power to protect or promote, harm or hinder, by magic, the individual or collective dairying process.

The essays in this book are the results of the Ninth International Conference on Ethnological Food Research hosted by the Department of Irish Folklore, University College Dublin, Ireland, in June 1992. The conference theme, 'Milk and Milk Products', was chosen in recognition of the historic and modern-day importance of these commodities in the Irish diet and economy. Liquid milk, for example, is still an essential part of the adult diet in Ireland, and the milkman, making his early morning door-to-door deliveries, is as familiar a figure as the postman. Although *per capita* consumption of milk has declined from a peak of 211 litres in 1984, it is still more than treble the EC average, placing Ireland fourth in the world table of milk-drinking countries behind Norway, Finland and Iceland. While the obvious importance of milk consumption in Ireland today can be paralleled in few other regions, yet milk, butter, cheese, skimmed milk, curds and whey are ancient basic foods in many parts of the world, and as such, they—and other basic foods—are worthy subjects for international scholarly research. These essays are a starting point for such an endeavour, and the select bibliographies of milk and milk products supplied by many of the authors in this volume are an important contribution towards further study of this branch of foodways scholarship.

Patricia Lysaght
Lá Fhéile Pádraig, 1993

Part I

National and Regional Surveys:
Everyday, Festive and Crisis Diet

1 The consumption of dairy products in the Netherlands in the fifteenth and sixteenth centuries

Johanna Maria van Winter

The Belgian nickname for the Dutch is *kaaskoppen*, which means 'cheese heads', a nickname that we might as well treat as a title of honour because of the high quality and the nutritious value of the dairy product to which we are thus compared. Even though we are not the only people to produce a variety of high-quality cheese, it is true that this dairy product must have played an important role in our culture from early times. There is, for instance, a tenth century hagiography of St. Adalbert van Egmond, which contains an anecdote about a farmer who stole a cheese which had been sacrificed to this saint. The farmer was punished when he tried to eat the stolen cheese: he bit off his own fingers. When he touched the saint's relics, however, he was miraculously healed. He was then presented as a serf to the Church of Egmond by his lord.[1]

I do not intend to go through all medieval sources concerning cheese and other dairy products in the Netherlands, but I have the impression that the consumption of these products, which was not inconsiderable as early as the fourteenth and fifteenth centuries, increased especially in the sixteenth century. Perhaps the Netherlands of those days was not yet the exporting country of dairy products that it is today, but it has been established that the domestic trade of these products must have grown rapidly since the second half of the sixteenth century.[2] In order to find support for the assumption that this was not only an increase in trade, but also in consumption, I will make use of housekeeping books[3] and especially of cookery books.[4]

Pure milk, that is, in its unpreserved form, was not an important ingredient in the medieval kitchen, because it was highly perishable, and secondly because it was not unconditionally considered healthy: watery milk was said to purge the bowels, and fat milk was said to restrict the veins. Milk should also not be combined with fish. Milk was believed to cause headaches, and because it was supposed to be bad for one's teeth, people were advised to rinse the mouth with honey-water after drinking it.[5] However, there was one traditional dish in which milk was used, namely *furmenty* or wheatmeal porridge, which was eaten with roast game. In several French and English cookery books this dish appears under names that are derived from the Latin word *frumentum*, which means wheat.[6] In the Dutch cookery book *Notabel boecxken van Cokeryen* from around 1510, this dish is discussed under the name 'Gharneye', and in Vorselman's *Nyeuwen Coock Boeck* from 1560, the same dish is mentioned in a nameless recipe in the section on Potages:[7] 'Boil hulled grains of wheat in water and drain them; next, boil plain milk with egg yolks, sugar and saffron. Add the boiled grains of wheat and let them boil for a while'.

Preserved forms of milk are butter and cheese, and these had the advantage of being less perishable, but just like milk, they were not considered healthy. Butter was said to make the food swim in the 'orificium' (the opening) of the stomach and to purge more than was becoming, although it was considered suitable for preparing food. Fresh cheese was hard to digest, caused gallstones, and 'did not suit the regimen of health. Briefly, no cheese is wholesome, except that which is eaten least of all'.[8]

Still, butter was a product that was used quite regularly in the Flemish and Dutch kitchens, whereas in other countries lard or oil was used. Foreigners mocked 'the Flemish' (a term that probably denoted all the inhabitants of the Low Countries) for their use of butter, both in the fifteenth and in the sixteenth centuries. In a historical song from 1470 set against the background of the Wars of the Roses, it is said that the Flemish are mere peasants, made of butter, and that the English are not used to eating butter in Flanders, but to drinking cider in Normandy.[9] A Latin treatise on food from 1560, written by a Frenchman, says in the chapter entitled 'De Butyro', that the Bretons preserve and export much butter, but that it is the Flemish who actually use it as a special foodstuff—they even put it in their drinks, and they use it every day in every meal, and everyone who visits Flanders should take along a knife in order not to be misled by elegant butter balls.[10]

As regards cheese, several kinds, made of sheep's milk as well as cow's were available on the Dutch market in the fifteenth and sixteenth centuries: *kanterkeese*[11] (cummin cheese), *harnescher*, *hernese* or *harvyskeese*,[12] and *caseum nymolken* (beestings cheese), as well as *case van Gauy* or *kase de Gein*[13] (new spring cheese), *platten keese*[14] (cottage cheese, sold in flat boxes made of split-wood), *Engelschen keese*[15] (a cheese that could be grated), *caseum Zelandinum*[16] (Zeeland cheese), *caseum ovinum*[17] (sheep's cheese), cheese from Brie[18], Parmesan cheese[19], cheese from Tienen and Nivelles[20] (firm cheese resembling English cheese), and also *vleckier*,[21] cheese the nature and substance of which can often, but not always, be deduced from the cookery books. In addition, numerous references can be found to hard or soft, white, fat or fresh cheese, of which no specific names are mentioned.

Furthermore, there were also curd cheese-like dishes that could be home made, with names like *moelgen*, *eyerkaes* or *present*,[22] *wronghel* and *liesen*,[23] and *recotten*, *rijskotten* and *kodt*,[24] derived from the Italian *ricotta* in the course of the sixteenth century. Another foreign name for *ricotta* was *cremijboellij* (cream boiled).[25] For this dish, the milk had to be made to curdle, so that the whey could be drained and a solid mass formed. When people wanted to make *Melck ghelardeert* (milk larded), they could add thinly cut bacon, press the mass into a cheese mould, and then cut it in slices and fry it in lard.[26] This method of preparation was also used in the countries surrounding the Netherlands (*lait lardé* or *leche lardys* and *ghespeckede melk*.[27] Another method was to homogenise the curd by rubbing it through a sieve and then adding butter, cream or rosewater and sugar.[28]

The curdling of the milk could be stimulated in various ways by boiling it with eggs, or adding buttermilk to plain milk, or sour cream to fresh cream were the most common methods in the Low Countries.[29] Twice I have found vinegar mentioned as a means of curdling hot milk.[30] Curdling by adding ale to hot milk, making a *poshotte* of ale,[31] is a method I have not found in Dutch cookery books. A very clever way, which avoided the risk of the milk getting burned, was to heat it in a *bain-marie*: a closed earthenware pot with milk and eggs was heated in a kettle with boiling water. The effect that we would aim for nowadays—a non-curdled smooth custard—was not achieved, because the water was kept boiling, not just heated to boiling point. Therefore, in the recipes concerned, the curdling process always involved draining whey.[32]

A remarkable Dutch variant is the *rutins kese*, made of naturally soured and curdled milk.[33] In addition, at the end of the sixteenth century, we come across a dish called *hangop* for the first time, a curds-like dish that is not prepared with milk but with buttermilk.[34] The word *oiste*, which occurs in a recipe from the middle of the sixteenth century, probably means *hangop* as well.[35]

In addition to curd cheese and derivatives of it, it is also possible to use milk in order to make custard or a *vlaede*. I have not found any examples of this in foreign cookery books, but I did find a large number of examples in Dutch books. This does not mean that the Dutch invented it. From references in foreign cookery books it appears that the *flaons* were known and even popular not only in the Netherlands, but also in other countries in the fourteenth and fifteenth centuries. However, no recipes are given, at best a Lent version, *flaons en karesme*, i.e. a recipe without eggs or dairy products.[36] Although this dish was sometimes, or usually—the Dutch recipes are not always clear on this point—baked in the oven, as a rule it did not have a pastry base. It was, however, usually mixed with flour, bread crumbs or other flour products, so that the whey thickened after being separated through heating.[37] The *gefruytte sane* or *cremefrijt* was based on the same principle, be it that it was always boiled. Subsequently, it could also serve as the filling of a pie.[38] The name of this recipe suggests a French origin, and indeed it occurs in the incunabulum edition of the

6 Milk and milk products

Viandier by Taillevent, from about 1490. I have not found it in other foreign cookery books from the Middle Ages.[39] In addition, the Dutch also made several kinds of cheese and cream pies with a pastry base.[40]

What was called *cremefrijt* in the Netherlands was actually a kind of porridge, or a milk pudding after it cooled down. The word pudding was still unknown in those days, but porridge recipes did begin to appear in the course of the sixteenth century, also with names such as Spanish porridge or German porridge.[41] However, I have not found any examples of these porridges in German or Spanish cookery books—perhaps they matched the French *flaons* in popularity, so that recipes were considered superfluous. These porridges were sometimes thickened with rice meal, and occasionally with wheat flour; eggs were also usually added and sugar was one of the standard ingredients.

Moreover, people started to prepare the time-honoured *blancmange* with cow's milk as well, whereas originally it was made of almond milk. Almond milk thickens when it is boiled. It is made from mashed almonds diluted with water or other liquids until it is a milky fluid. If it is replaced by ordinary milk, which does not have this quality, a thickener such as rice meal has to be added. In a blancmange, which people also began to prepare with ordinary milk in the sixteenth century version, it was customary to boil cubes of well-cooked chicken; this gave a blancmange of Spain, a porridge of milk and rice meal, sugar and chicken, also simply called a *blancmange*.[42] If the chicken was left out, and egg yolks added, the dish was called a *blancmange* in Lent,[43] which is in fact a very strange name, for during Lent not only chicken but milk and eggs as well were forbidden. Perhaps the reference was not to the actual Lent—the forty days before Easter plus a few occasional so-called 'Ember days'—but to the days of abstinence or fast days, occurring twice a week. On those days, people were allowed to consume dairy products and eggs, but not meat. The Reformation was to put an end to these fasting and abstinence precepts, but in the first half of the sixteenth century the Reformation hardly played a part in the Netherlands.

Reformation or not, my impression is that the dairy consumption in the Netherlands increased enormously in the sixteenth century. Is this right or is it just an illusion, caused by the fact that Dutch cookery books did not appear until the end of the fifteenth century, and increased in number in the sixteenth century? In other words, is it a question of the more recipes there are, the more dairy recipes you will find?

An argument against this view is, in my opinion, the multitude of new names for an already existing product, like curdled milk, from which the whey was drained. Of course, people had already been familiar with this process for a long time: they made dishes using it, to which they gave French names: such as *lait lardé* and *cremebouillie*, or the literal Dutch translation *melck ghelardeert* and *cremyboelly of sanen*. But in addition to these two foreign names, as many as eight other names were used in the Netherlands towards the end of the fifteenth century and in the sixteenth century: *moelgen, eijerkaes, present, wronghel, liesen, recotten/rijskotten/kodt, rutins kese* and *soet melck*. Furthermore, people were familiar with at least three methods of curdling milk: by boiling the milk with eggs, in a bain-marie if so desired, by adding sour milk to the hot milk, or by adding vinegar to the hot milk. The English 'poshotte of ale' was unknown in the Netherlands, but if it had been known, it would have been the fourth method of curdling milk.

In my opinion, the large number of custard recipes also gives support to the idea of a growing demand for the processing of plain milk, just like the 'porridge' recipes and the preparation of a traditional dish like *blancmange* with cow's milk instead of almond milk. In addition, there were the many recipes for cheese and cream pies, for which a whole range of new and fully matured, soft and firm cheese, appears to have been available.

When we compare these particulars with the results of agronomico-historical research, they agree wonderfully well. Although there are hardly any contemporary descriptions of the Dutch livestock husbandry in the sixteenth century, we can nevertheless deduce from cattle counts and indirect information that, as a consequence of specialisation in this branch of industry, improvements in cattle feed and manuring of pastures, the livestock industry in Flanders, Holland and Friesland, grew enormously between the beginning of the sixteenth century and the middle of the seventeenth century. Furthermore, in Friesland around 1570, the milk yield per cow averaged at least 1350 litres a year, and in the province of North Holland around 1640, it averaged already more than 2000 litres a year, whereas in Denmark about 1800, the regular milk yield per cow was approximately 600 litres a year.[44] The inventory of butter and cheese dairies in a small Frisian hamlet testifies to this. The brass cheese kettle in which the whey was heated was the most valuable part of the inventory:

> In Hennaarderadeel in 1554–62 only two-thirds of the farmers with at least 10 milk cows owned a cheese kettle. The average capacity of those in use was just over 0.50 *ton*. In 1595–1600 all large farmers possessed cheese kettles and their average size was now 0.75 *ton*; by 1646–1654 the average had risen considerably to 1.25 *ton*.[45]

In conclusion, we may say that the growth and quality-improvement of stock breeding in the border provinces of the Netherlands—Flanders, Holland and Friesland—in the course of the sixteenth century, led to a marked increase in the supply of plain milk, butter and cheese, which were sold primarily on the domestic market. The consumer actually benefited from the possibilities provided by this supply, which is reflected in cookery books. The numerous ways of processing plain milk and even buttermilk to make curd cheese and *hangop*, as well as the numerous new names that were introduced for these products, countless recipes for thickened *vladen*

8 *Milk and milk products*

(custards or flans) prepared with milk and eggs, and recipes for cheese and cream pies, bear testimony to this.

NOTES AND REFERENCES

1 'De Vita Sancti Adalberti Confessoris', introduced, edited and translated by G.N.M. Vis, in *Egmond en Berne. Twee verhalende historische bronnen uit de Middeleeuwen*, 's-Gravenhage, Nederlands Historisch Genootschap, 1987 (Nederlandse Historische Bronnen VII), pp. 54–55, caput 11.
2 Jan de Vries, *The Dutch Rural Economy in the Golden Age, 1500–1700*, (New Haven and London, 1974), pp. 156–162, 166–167.
3 The most important sources are:

Lobith 1426/27 & 1427/28 = Accounts of the customs officer at Lobith for the years 1426/1427 and 1427/1428, unpublished, State Archives of Gelderland in Arnhem, Hertogelijk Archief (Ducal Archives) section A no. 729;

Lobith 1428/29 = Item for the year 1428/1429, same archives, section A no. 730; edited in *Het Keuckenboeck van de tollenaar van Lobith 1428/1429*, ed. G.R. Bosscha Erdbrink, Fontes Minores Medii Aevi 18, Groningen 1979;

Egmond = *Egmondse kloosterrekeningen uit de XIVe eeuw*, ed. J. Hof, Fontes Minores Medii Aevi 17, Groningen 1976;

Tzwynnen = Account book of Mr Floris Tzwynnen, 1493/1509, unpublished, Municipal Archives Utrecht, Stad I no. 818, 'Rekenboek van Mr. Floris Tzwynnen'. For this source see: Gisela Gerritsen-Geywitz, 'Drie recepten uit het *Rekenboek van Floris Tzwynnen*, in *Een school spierinkjes. Kleine opstellen over Middelnederlandse artes-literatuur*, Hilversum 1991, pp. 72–76.

4 The following Dutch cookery books (more or less in chronological order) will be referred to as follows:

Serrure = C.A.S[errure] (ed.), *Keukenboek uitgegeven naar een handschrift der vijftiende eeuw*, Gent, 1872 (Maatschappij der Vlaamsche Bibliofielen, Werken nr. X); ms. Gent UB nr. 1035; 62 recipes.

Gent 15−A = W.L. Braekman (ed.), 'Een belangrijke Middelnederlandse bron voor Vorselmans 'Nyeuwen Coock Boeck (1560)', in *Volkskunde* 87 (1986), pp. 1–24; Gent 15−B+C = W.L. Braekman (ed.), *Een nieuw Zuidnederlands Kookboek uit de vijftiende eeuw*, Brussel, 1986 (Scripta, Mediaeval and Renaissance Texts and Studies, 17); ms Gent, Koninklijke Academie voor Nederlandse Taal-en Letterkunde, no.15; Gent 15−A+B+C together contain 430 recipes.

VdNoot = *Een notabel boecxken van cokeryen*, Brussel, Thomas van der Noot, *c.* 1510; Facsimile edition, under the title *Het eerste Nederlandsche gedrukte kookboek* (The first Dutch cookery book in print), 's-Gravenhage, 1925; 170 recipes.

Gent 476 = Ria Jansen-Sieben en Johanna Maria van Winter (eds.), *De keuken van de late Middeleeuwen*, Amsterdam, 1989; ms Gent UB no. 476; 267 recipes.

Vorselman = Elly Cockx-Indestege (ed.), *Eenen Nyeuwen Coock Boeck, Kookboek samengesteld door Gheeraert Vorselman en gedrukt te Antwerpen in 1560*, Wiesbaden, 1971; 496 recipes.

Wirtsung = [Karel Baten], *Eenen seer schoonen ende excelenten Coc-boeck*, appendix to Christophorus Wirtsung, *Medecyn-Boec*, 2 edn., Dordrecht, Jan Canin, 1593; 293 recipes.

Antwerps = J. Lindemans (ed.), 'Een Antwerps receptenboekje van circa 1575–1625', in *Verslagen en Mededelingen van de Koninklijke Vlaamse Academie voor Taal- en Letterkunde*, 1960, pp. 401–434; 85 recipes.

The following foreign cookery books will be referred to:

Menagier = *Le Menagier de Paris*, ed. Georgine E. Brereton and Janet M.Ferrier, Oxford 1981.

Taillevent = *Le Viandier de Guillaume Tirel dit Taillevent*, ed. Jérôme Pichon et Georges Vicaire, nouvelle édition par Sylvie Martinet, Genève, Slatkine reprints, 1967.

Viandier = *The Viandier of Taillevent*. An edition of all extant Manuscripts, ed. Terence Scully, University of Ottawa Press 1988.

Fifteenth century = *Two Fifteenth Century Cookery-Books*, ed. Thomas Austin, published for the Early English Text Society by the Oxford University Press, London—New York—Toronto, 1964.

Curye = *Curye on Inglysch*, English culinary manuscripts of the fourteenth century (including the *Forme of Cury*), ed. Constance B. Hieatt and Sharon Butler, published for the Early English Text Society by the Oxford University Press, London—New York—Toronto, 1985.

Pottage = *An Ordinance of Pottage*. An edition of the fifteenth century culinary recipes in Yale University's Ms Beinecke 163, ed. Constance Hieatt, London, Prospect Books, 1988.

Martino = 'Libro de Arte Coquinaria, composto per lo egregio Maestro Martino', in *Arte della Cucina*, Libri di ricette, testi sopra lo scalco, il trinciante e i vini, a cura di Emilio Faccioli, Milano 1966, pp. 115–204.

Kochbuch = H. Wiswe, 'Ein mittelniederdeutsches Kochbuch des 15. Jahrhunderts', in *Braunschweigisches Jahrbuch* 37 (1956) pp.19–55, & *Idem*, 'Nachlese zum ältesten mittelniederdeutschen Kochbuch', in Item 39 (1958) pp. 103–121.

5 *Tregement der ghesontheyt*, Brussel, Thomas van der Noot, 1514, folio M II verso/M III recto, caput xciii (no modern edition).

6 E.g. *Fourmentee*: *Menagier*, p. 246 nr. 234; *Fromantée*: *Taillevent*, pp. 188/189; 'Furmenty with venyson': *Fifteenth century*, p.70; *Frumente* and *Frumente*: *Curye*, pp. 62 and 98.

7 VdNoot, folio F I recto; 'Van Pottagien', Vorselman p.113.

8 *Tregement der ghesontheyt*, folio M III recto.

9 A.J.V. le Roux de Lincy, *Chants historiques et populaires du temps de Charles VII et Louis XI*, Paris 1857, p.157, 'Ballade aux Franchoix', stanza V, and p.161, 'Response sur la dicte ballade', stanza VIII.

10 Joannes Baptista Bruyerinus Campegius, *De re cibaria libri XXII*, Lugduni [Lyon] 1560, p.754.

11 Lobith 1426/27 fol. 40 verso, 5 *waag* of cheese purchased, consisting of 90 *kanter* cheeses. A *waag* is a unit of weight. One *waag* of cheese weighed about 180 pounds.

12 Egmond p.108–109, account-year 1387/88, purchase of a *waag* of *Harmersche* or *Harnescher* cheese. Cf. A.A. Arkenbout, 'Het jaar round op de Hof te St. Maartensdijk (1457–1458)', in *Castellogica, Verkenningen*; Mededelingen van de Nederlandse Kastelen Stichting I (1986–1987), p. 250: e.g., during the week of 13–20 August 1457 in the household of the lord of St. Maartensdijk in the province of Zeeland the following dairy products were consumed: 8 cheeses from the *joncwijfs huysse* (the house of the dairy maid—so presumably ripened cheeses), 2 small *hernesse* cheeses (young, unripened cheeses), 1/2 barrel of butter. Gent 15–B+C, p. 40 nr 33: to make cheese rissoles in the Lombardian way ... take three eggs, and a round *harvyskeese*, and put these together in a dish ... Contrary to what the editor of this text, W.L. Braekman in his glossary p.135 suggests, viz. that *harvyskeese* is cheese from the place Herve, I think that the word *harvyskeese* is an incorrect spelling of *harnyskeese*, i.e. *hernesse keese*.

13 Tzwynnen p.142, Anno 1500: 'Prior Vlissingensis promisit 'caseum nymolken.' Serrure p.6 no. XXI: 'case van Gauy'; Serrure p.10 no. II: 'kase de Gein'. Cf. *Viandier*, glossary and index p. 351: gain 'herbe du pâturage, regain, moisson,

10 Milk and milk products

époque de la récolte'—fromage de gain, creamy cheese from the milk of pastured cows, goats.

14 Gent 15−B+C p. 48, no. 75: *platten keese* (cottage cheese); VdNoot folio E II recto: *platten case*; Gent 476 p. 98 no. 98: *plantenkeese*. *Plantenkeese* (herb cheese), is probably a wrong spelling for *platten keese* (cottage cheese), as on p. 58 in the index to the ms. Wirtsung folio A III verso [no.67]: *platte Keese*; Wirtsung folio B II verso [no. 217 and 218]: *platten Keese van Botermelc* and *platte Keese van Soetemelck* (cottage cheese of buttermilk, and cottage cheese of plain milk).

15 Tzwynnen p. 144, Anno 1504: 'Prior in Vlissingen misit cum Dyonisio Jacobi unum caseum anglicanum. Item frater quidam dedit 1 caseum zelandicum.' Gent 15−B+C, p. 104 no. 268: to make cheese broth, take cheese from Tienen, thinly sliced English cheese and the like, and grate the cheese or cut it into slices ... Gent 15−B+C, p. 112 no. 292: to bake cheese cakes, take English cheese, grate it finely ... VdNoot folio E IV recto/verso: to make cheese cakes of potherb, take English cheese that is very finely grated or finely pulverized ...

16 Tzwynnen, see note 15.

17 Tzwynnen pp. 60, 155, 171, 174, 180.

18 Vorselman p. 203, XIV no. 1: 'kese van Bry'; XIV no. 2: 'kese van Brij'.

19 Vorselman p. 214, XIV no. 46: 'Parmesaens kees';

20 Gent 15−B+C, p. 104 no. 268, 'Tiensche keese'; Gent 476, p. 123 no. 177: 'Nievelsche keesse' and 'Tijnse kees'; Gent 476, p. 123 no. 178: 'Nievelsche kees'; Vorselman p. 195, XII no. 2: 'Thiensen kese'; Vorselman p.199, XIII no. 6: 'Thiensen kase'.

21 Gent 476 p. 68, no. 22: 'to make soaked bread in the Jacobean way, take a roast chicken and bone it and take *vleckier* or another good cheese and cut it into thin slices ...'

22 Serrure p. 15 no. XXV: 'moelgen'; Gent 15−A, p. 21, no. 76: 'eijerkaas' (egg cheese), and no. 77: 'present'. The same recipes can be found in Vorselman p. 208, no. 21 and no. 22, but in reverse order. Gent 476, p. 154 no. 262: 'eijerkese'.

23 Gent 15−A, p. 21, no. 76: 'wronghel' (curd); Gent 15−B+C, p. 125 no. 337: 'liesen'; Gent 476, p. 107 no. 120: 'wrongel'; Gent 476, p. 107 no. 121: 'wrongel'; Gent 476, p. 154 no. 263: 'vrongelen'; Gent 476, p. 154 no. 264: 'vrongelen'; Vorselman p.208, no.XIV 22: 'wronghelen'.

24 Gent 476, p. 154 no. 263: 'kodt'; Antwerps, p. 406 no.10: 'Recotten'; Antwerps, p.409 no.20: 'Rijskotten'.

25 Gent 476 p. 99, no. 99: 'cremijboellij'; Wirtsung folio A III verso [no. 69]: 'Cremyboelly of Sanen' (boiled cream or double cream). I have not found a French example, but I have found an English one: *Pottage* p. 60 no. 76: 'Creme boyled', although in this recipe, the hot milk is curdled by adding egg yolks, whereas in the Dutch recipes it is done by adding sour cream. *Fifteenth century* p.8 no.xiii: 'Creme Boylede' is not a recipe for curdled milk but for thickened custard: the Dutch name for this recipe could be *cremefrijt* (see below, note 38).

26 VdNoot folio D I recto, and Vorselman p. 210 no. XIV 34: *Melck ghelardeert* (larded milk).

27 *Lait lardé*: Menagier, p. 254 no. 259; *Layt lardé*: Taillevent p. 123, and *Viandier* p.254 no. 200; *Lait lardé* moreover in the incunabulum edition: Taillevent p.159; *Lette lardes* and *letlardes*: Fifteenth century p. 35 no.viii and p. 92; *Lete lardes*: Curye p. 113 no. 69; 'To make leche lardys of iii colours': *Pottage* p. 42 no. 19; *ghespeckede melk*: Kochbuch pp. 41−42 no. 61, 62.

28 Gent 15−A, p. 21 no. 77: curd cheese with butter and sugar; Gent 15−B+C, p. 125 no. 337: curd cheese with cream and sugar; Vorselman p.208 no. XIV 21: curd cheese with butter and sugar; Gent 476 p. 142 no. 236: curd cheese with butter and sugar; Gent 476 p. 154 no. 263: curd cheese with egg yolks,

cream and sugar; Gent 476 p. 154 no. 264: curd cheese with egg yolks, rose water and sugar; Antwerps p. 406 no. 10: curd cheese with cream and sugar; Antwerps p. 409 no. 20: curd cheese with egg yolks, rose water and sugar.
29 Serrure p. 15 no. XXV: boil sheep's milk with egg yolks in a bain-marie, and drain the whey through little holes in the eartherware pot; Gent 15–A, p.21 no. 76: boil; milk with eggs in a bain-marie, and drain the curds; Gent 15–B+C, p.109 no. 283: boil cream with eggs in a bain-marie until it is stiff, put it in a clean cloth and press it into a cheese mould; Gent 15–B+C, p. 124 no. 335: boil plain milk and cream with eggs in a bain-marie until it curdles, drain it in a cloth and press into a small mould; Gent 15–B+C, p. 125 no. 337: boil sweet cream, cool it until it is tepid. Add sour cream, drain the whey and press the curd cheese into a mould; Vorselman p. 208 no. 22: boil the milk with eggs in a bain-marie and drain the whey; Gent 476, p. 99 no. 99: boil plain milk, cool it until it is tepid. Add sour cream and let it set and curdle on a warm spot for one night; Gent 476, p.140 no. 232: bring plain milk with eggs and buttermilk to the boil, let it thicken and curdle, and drain the whey; Wirtsung folio A III verso [no. 69]: boil plain milk, cool until tepid and add sour cream. Let it stiffen and curdle on a warm spot for one night; Antwerps p. 406 no. 10: bring plain milk with rose water to the boil and add buttermilk. Drain the curd cheese; Antwerps p.409 no. 19: (boil cream and mix it with as much buttermilk, drain in a cloth and mix with egg yolks, sugar and rose water; Antwerps p. 409 no.20: 'bring plain milk to the boil and add buttermilk. Drain the curd cheese, mash it and add egg yolks and rose water; drain once more.'
30 Gent 476, p.142 no.236: to make *soet melck*, take plain milk and heat it. Add a little vinegar that curdles, then hang in a cloth and drain the whey. Gent 476, p. 154 no. 263: To make *kodt*, take cream, put it on the fire in a clean pot and when it almost starts to boil, take two spoonfuls of elderberry vinegar (*fliereeck*) and then take it from the fire . . .
31 E.g. *Fifteenth century* p. 36 no. x: 'Vyaund de leche.-; & then make a styf poshote of Ale; p. 36 no.xi: 'Vyaund leche.-, & make a styf poshotte of Ale; p. 36–37 no.xii: 'Vyaund leche.- & with ale make it to a poshotte; then hange the croddys in a pynne, & let it ouer-renne;
32 See the examples in note 29.
33 Gent 476, p. 154 no. 264: to make *rutins* cheese, take one mingelen of *rutin* and put it in a clean cloth and let it hang until all the whey has run out.
34 Wirtsung folio A III verso, no. 68: to make a buttermilk cheese, take thick buttermilk, and hang it in a cloth to drain the whey; then take the thick substance, and sweet cream, egg yolks and sugar.
35 Gent 476, p. 140 no. 232: to make a curd cake, take plain milk and bring it to the boil and then take an earthenware dish full of buttermilk, but you must take the thickest part of the buttermilk and of the *oiste*. This cookery book from Gent was written by eight different hands in the period between the beginning of the sixteenth century and 1584. This recipe was written by hand E, around the middle of the century.
36 E.g. *Menagier* p. 249 no. 247: 'Flaons en Karesme'; *Viandier* p. 214 no. 152: 'Flans, tartes en Karesme'; *Taillevent* p. 31 nos. 108, 222: 'Pour faire flaons et tartes en Quaresme'.
37 Serrure p. 15 no. XXVI: *Ghesoden vladen* (boiled custard); Gent 15–B+C, p.48 no. 73: *Om zaenvlaeden*' (cream custard); Gent 15–B+C, p. 48 no. 74: *Om appelvlaeden* (apple custard); Gent 15–B+C, p. 51 no. 84: *Item eyervladen te backen* (egg custard); Gent 15–B+C, p. 125 no. 336: *vlaijen van sanen* (custard of cream); VdNoot folio E II verso: *Om saen vladen te maken* (to make cream custard); VdNoot folio E II verso: *Noch om vladen* (to make other custards); Gent 476 p. 106 no. 119: *sonderlynge vlade* (special custard); Gent 476 p. 110

no. 132: *Om vlaijen te maken* (to make custard); Vorselman p. 204 no. XIV 3: *Ghemeyn vladen* (ordinary custard); Vorselman p. 204 no. XIV 4: *Om groen vladen te maken* (to make green custard); Vorselman p. 214 no. XIV 48: *Een vlade aent vier oft panneken genaemt diriola* (a custard called *diriola*); Item no. XIV 49: *Een diriola buten der vasten* (a *diriola*, not to be eaten during Lent); (In the Italian recipes by Maestro Martino and Platina, which were the examples for Vorselman's recipes, the filling is baked in a pastry crust: Martino p. 172 *Diriola*, and p. 173 *Per farne una in Quadragesimo*.) Wirtsung folio A II verso [no. 45]: *Een Melck-Vlade te maken* (to make milk custard); Item [no. 47]: *Om een Quede Vlade te maken* (to make Quince custard); Item folio A II verso/A III recto [no. 48]: *Om een sonderlinge Vlade te maken* (see above, Gent 476 p. 106 no. 119). Item folio B II verso [no. 216]: *Om een Vlay in een schotel te backen* (to bake custard in a dish); Antwerps p.408 no. 17: *Om vlayen te maecken* (to make custard); Antwerps p. 409 no. 18: *Om een witte vlaye te maecken* (to make a white custard).

38 VdNoot folio F I verso: cream boiled with crumbs of white bread or of rolled wafers, and egg yolks; Gent 476 p. 81 no. 52: plain milk boiled with white breadcrumbs, egg yolks and butter; Vorselman p. 201 no. XIII 15: here, the boiled mixture of milk, breadcrumbs and eggs is then used as filling for a cream cake; Vorselman p. 202 no. XIII 18: cream boiled with crumbs of white bread or rolled wafers, and egg yolks; Wirtsung folio A II verso no. 42: here, the boiled mixture of cream, butter, egg yolks, and white breadcrumbs can be used as filling for a pie that is baked in the oven. All these recipes contain much sugar and little salt.

39 *Taillevent* p. 160 (the incunabulum edition of the Viandier): *Pour cresme fricte*.

40 Serrure p. 6 no. XXI: dough to make 'pipesen'; Item p.10 no.II: pasties in dishes, made of cheese *de gain*; cf. note 13; Gent 15−B+C p. 50 no. 80: to make a cream pie; VdNoot folio E II recto: to make brown pies; Item: when you want to make cheese pies of soft cheese; Item: *lecfriten oft gouwieren*; Gent 476 p. 88 no. 71: to make cheese pies; Item p. 123 no. 177: *lackfric spijse*; Item p. 123 no. 178: *flamijtschen*; Item p. 123 no. 179: a *lackfric* dish of cheese pies; Item p. 140 no. 232: to make a curd cake; Vorselman p. 199 no. XIII 5: to make a brown pie; Item p. 199 no. XIII 6: to make a cheese pie; Item p. 199 no. XIII 7: a pie of fresh cheese; Item p. 201 no. XIII 16: a pie with peeled boiled eggs; Item p. 202 no. XIII 20: pies of soft cheese; Item p. 202 no. XIII 21: *lecfrijten oft gouwieren*; Item p. 203 no. XIV 2: a cheese pie; Wirtsung folio A II recto [no. 30]: to make a cheese pie; Item folio B III recto [no. 223]: cheese pie; Antwerps p. 404 no. 2: to make a white pie; Item p. 405 no. 6: to make a pie of Spanish porridge; Item p. 405 no. 7: to make an English pie; Item p. 405 no. 8: to make a cream pie; Item p. 406 no. 9: to make a rice pie.

41 Gent 476 p. 98 no. 96: porridge for two dishes; Gent 476 p. 141 no. 234: for German porridge, make little balls of wheat flour, eggs and hot milk, and let them sink into the boiling milk or cream, so that 'plain milk with lumps' (*zoetemelk met brokken*) is formed, a name also known from a Dutch children's song; Gent 476 p. 143 no. 240: porridge from milk. Wirtsung folio A I recto [no. 6]: porridge or *crême de Moerbeck* in wintertime; Wirtsung folio A I recto [no. 7]: German porridge, see above, milk with lumps; Wirtsung folio A I recto [no. 8]: elderberry porridge; Wirtsung folio A I recto [no.9]: a special porridge; Wirtsung folio B II recto [no. 205]: rice in the Antwerp way; Wirtsung folio B II recto [no. 206]: porridge from rice meal; Wirtsung folio B II recto [no. 207]: Spanish porridge; Antwerps p. 408 no. 16: Spanish porridge. Sugar is added to all these porridges.

42 Gent 476 p. 124 no. 181: blancmange the Spanish way; Wirtsung folio A I recto [no. 5]: blancmange of Spain; Antwerps p. 407 no. 12: blancmange. In all three recipes the fresh milk is boiled with rice meal, after which the meat of a boiled chicken or capon, as well as sugar is added and boiled with it.

43 Gent 476 p. 148 no. 249, & Wirtsung folio A I recto [no.2]: blancmange in Lent.
44 Jan de Vries, *Dutch Rural Economy* (see above, note 2) pp. 137–144.
45 *Ibid,* p. 144.

SELECT SOURCES AND BIBLIOGRAPHY

Sources

C.A.S [errure] (ed.), *Keukenboek uitgegeven naar een handschrift der vijftiende eeuw*. Gent, 1872 (Maatschappij der Vlaamsche Bibliogielen, Werken nr. X); ms. Gent UB nr. 1035.

W.L. Braekman (ed), 'Een belangrijke Middelnederlandse bron voor Vorselmans 'Nyeuwen Coock Boeck (1560)'. In *Volkskunde* 87 (1986), pp. 1–24.

W.L. Braekman (ed.), *Een nieuw Zuidnederlands Kookboek uit de vijftiende eeuw*. Brussel, 1986 (Scripta, Mediaeval and Renaissance Texts and Studies, 17); ms Gent, Koninklijke Academie voor Nederlandse Taal- en Letterkunde, no. 15.

Een notabel boecxken van cokeryen, Brussel, Thomas van der Noot, c. 1510; Facsimile edition, under the title *Het eerste Nederlandsche gedrukte kookboek* (The first Dutch cookery book in print). 's-Gravenhage, 1925.

R. Jansen-Sieben en J. M. van Winter (eds.), *De keuken van de late Middeleeuwen*, Amsterdam, 1989; ms Gent UB no. 476.

E. Cockx-Indestege (ed.), *Eenen Nyeuwen Coock Boeck, Kookboek samengesteld door Gheeraert Vorselman en gedrukt te Antwerpen in 1560*. Wiesbaden, 1971.

[Karel Baten], *Eenen seer schoonen ende excelenten Coc-boeck*, appendix to Christophorus Wirtsung, *Medecyn-Boec*, second edition, Dordrecht, Jan Canin, 1593.

J. Lindemans (ed.) 'Een Antwerps receptenboekje van circa 1575–1625', in *Verslagen en Mededelingen van de Koninklijke Vlaamse Academie voor Taal-en Letterkunde*, 1960, 401–434.

Bibliography

Jan de Vries, *The Dutch Rural Economy in the Golden Age, 1500–1700*, New Haven and London, 1974, 156–162, 166–167.

2 Milky ways on milk-days: the use of milk products in Hungarian foodways

Eszter Kisbán

From the late fifteenth century until the middle of the nineteenth century, live cattle were the export product through which Hungary joined in the European economy. Dairy products, such as cheese and butter, did not play a role even in domestic interregional trade. They were made primarily for family use, and they reached the local markets only in small quantities. When certain Hungarian writers speak of the 'prime importance' of dairy products in Hungarian food, they do so without having made any comparisons with Europe's traditional and modern foodways. In reality, the consumption of milk and dairy products in Hungary, both historically and in modern times, is rather low. In the middle of the seventeenth century, the economic historian László Makkai attempted to express in statistical terms the food consumption of peasants. In his model he put the annual average per capita consumption of milk at 155 litres, which included butter, curds, and cheese as well. Landless agricultural labourers were not even included in that model.[1] The first Hungarian food statistics, from the early 1880s, put the figure much lower in regard to the total population (see Table 2.1).

Using a contemporary key, the national average per capita consumption of milk and dairy products, expressed in milk, is 75 to 80 litres, while the corresponding figure for the towns and cities is 95 to 100 litres, of milk.

Unfortunately, twentieth century statistics do not consistently cover the particular dairy products separately (see Table 2.2).

Table 2.1 Per capita average consumption of milk and dairy products in kg. 1884[2]

	Hungary	The towns and cities of Hungary
butter	1.24	3.07
milk	31.92	47.93
curds, cheese	5.67	4.93

Table 2.2 Per capita average consumption of milk and dairy products in Hungary in kg.[3]

1934–38	1955–59	1960	1970	1980	1990
(a) milk and dairy products in aggregate, expressed in terms of milk (without butter):					
101.9	102.3	114.0	109.6	166.1	180.0
(b) butter					
1.0	1.1	1.4	2.1	2.0	2.0

Partial data indicate that 70 to 75 per cent of the milk was consumed in liquid unpreserved form, while the rest was consumed in its processed form.

THE MAIN HISTORICAL FORMS OF MILK PROCESSING

Prior to their settlement in the Carpathian Basin (896 AD), the Hungarians were a cattle-and horse-breeding people, with small-scale agriculture on the South Russian Steppe. Cattle-breeding was combined with milk consumption, the making of butter, curds, and what was called cheese (but was really dried curds), and probably yoghurt as well. In those days, the making of cheese using rennet was not a part of milk processing. The Hungarian words for curds—*túró*—and for dried curds—*sajt*—are Old Turkish loan-words from that period. In modern usage, *túró* means curds, while *sajt* denotes cheese made with rennet, ripened in the block and eaten from the block. In this paper, I shall, in accordance with modern usage, apply the term 'cheese' only to the product made with rennet and eaten from the block. In the Carpathian Basin in the Middle Ages, complex agricultural livestock-breeding economies emerged, with relatively low milk consumption. The principal milk-yielding animal was the cow, but sheep were also milked. In the surrounding regions in the south and west, cheese proper was part of milk processing. In Hungary also, cheese-making using rennet became a part of the processing of cow's milk, from at least the early seventeenth century through the influences of foreign (Swiss and Italian) dairy specialists employed on particular seignorial domestic farms. However, no new term was coined to denote cheese made with rennet: it, too, was referred to by the word *sajt*, just like preserved curds. For this reason, when one is reading sources from early modern times, it is not easy to say which kind of cheese is meant. There are small cheese-presses to be found as far back as the sixteenth century. Raising cattle for export was conducted in a special economic organisation, and from the thirteenth to the fifteenth centuries, proceeding from the southeast towards the northwest, a sheep-breeding pastoral population settled along the high-altitude range of the Carpathians.

16 Milk and milk products

The early groups were Wallachs of Balkan origin, but later some local ethnic groups also adopted this sheep-breeding culture. Their main dairy product was cheese made with rennet, which, however, they stored not in the form of cheese, but as curds, kneaded together with salt and kept in hermetically sealed vessels. As the names of its by-products show, this milk-processing technique did have an influence on the Carpathian Basin also.

In the sixteenth century, throughout the country, there were cases where butter, and so-called cheese from cow's milk, formed a regular part of the peasants' yearly payment to the landowners. Mediterranean-based Latin terminology was not consistent as regards the Hungarian situation; hence there is uncertainty as to the product called 'cheese' (*caseus*) in the records; was it real cheese made with rennet or dried curds? In any case, the cheese was counted by the piece, which means it was not salted curds packed into vessels. In numerous cases, the *caseus* had to be given to the landowners for a specified date in spring (at Easter, St. George's Day on 24 April, and Whitsuntide). If the quality was everywhere identical, the designated price-exchange values testify to cheese of various sizes. There is mention of cheese costing one, three, five, eight and ten denarii. (At the same time, one capon cost eight denarii.). Parallel with *caseus*, the sources also employed the term *formagia*, which is an Italianized form of Late Latin *formaticus (formaggio)*. *Formagia*, however, occurred far more rarely than *caseus*. If used with precision, the designation referred to cheese formed not with the bare hand but by means of a shape. In several regions where, in the sixteenth century, deliveries of *caseus/formagia* were exacted, peasants did not make cheese with rennet at all in modern times, but of course, it is also possible that cheese-making may have existed earlier in these regions, but had gone out of use with the passage of time.

In Austria, cheese was the staple dairy product in the Middle Ages. With the introduction of the plunge-churn, butter became the most important milk-derived product by early modern times. The situation in Alpine Austria, however, cannot be directly extrapolated to the Hungarian development. It seems more plausible to suppose that in Hungary in the late Middle Ages and early modern times, the peasants' cheese made from cow's milk consisted of cakes or balls of curds made without rennet, mostly formed with the hand or sometimes with a mould.[4] There is still another group of sources which have yet to be subjected to systematic analysis, namely, the price-lists for local markets written in Hungarian, which could furnish data on the distribution of 'curds' and 'cheese' sold by peasants in early modern times.

The earliest hint of agricultural improvement in Hungary was the setting up of new-style dairies on the estates. The first isolated example in 1680 inspired numerous similar enterprises in the late eighteenth century. These imported dairy-breeds of cows and specialist dairy-men from the West. The latter came chiefly from Switzerland and Austria, along with their families, but common parlance referred to all of them as 'the Swiss', and their work, too, was called 'Swiss dairying'. They were tenants on the estates, tending between fifteen and thirty cows of the estate and making butter, and also cheese with rennet. Here, then, was a second source of stimulation for the peasants to introduce cheese-making. But clearly the exercise had no impact. 'Swiss dairy-farming' tended to be concentrated primarily in the western portion of the country, the very region where, in the late eighteenth to the twentieth centuries, peasants did not make cheese with rennet.[5]

In Hungary, the overall transition from the old breed of cattle to the new dairy breed took place between 1880 and 1920. Following the development of late large-scale industrialization there was consequent growth of cities, and estates started in the 1880s to organize dairies to supply these. Farmers' co-operative dairies followed, and soon came to make use of the separator. This then led to the emergence of the modern milk industry.

From the eighteenth to the twentieth centuries, identifying the dairy products of peasants is no longer a problem. There were four different lines of milk processing. Here, however, I shall give only a brief outline of them, without mentioning all the by-products.[6]

1 *A practice known nearly all over the Carpathian Basin, and based mainly on cow's milk*, was to let full milk turn sour; this was then drunk as sour milk, or separated into sour cream and skimmed sour milk. The cream was used for cooking or made into butter. Skimmed sour milk was turned into curds by heating. Curds were used fresh, matured, dried or smoked. All were eaten with bread and porridges, pasta or in pies. Butter was melted (not salted) and stored. It was used for cooking only, never on bread.

2 *In south-east Transylvania*, instead of the above procedure, full cow's milk was left to turn sour in the churn; it was not skimmed, but some of its whey was drained off. It was then churned to homogeneous sour milk and so used, or else butter was made from it. No curds were needed since these were made from sheep's milk, in the following manner:

3 *Characteristic especially along the Carpathians, but not unknown in the Lowlands*, was the making of curds from sheep's milk, while cow's milk was also used regionally or sporadically. Fresh milk was curdled with rennet to make soft cheese. There were further products from the whey, including lean curds. The cheese was seldom used fresh in the block, but was regularly kneaded with salt and packed into storage vessels. This is the main product which was eaten with bread, porridges and pasta. In the Lowland areas, cheese from cow's milk was also matured in the block and sold as cheese to the non-agricultural population at the local markets.

4 *In the Lowland area, mainly amongst herdsmen*, yoghurt-processing was a regular practice (but not under that name). Farmers' wives might turn the yoghurt into curds as well.

Of these processing methods, numbers one and four are old Hungarian methods, number three is a Mediterranean method, and number two had a background in the Balkans.

MILK PRODUCTS IN THE HIERARCHY OF MEALS

That milk is a healthy nutriment is a modern perception. Earlier, the integration of dairy products into the meals system followed different principles. The first review provided by the sources of the role of dairy products in meals relates to the upper class culture of early modern times. Until the end of the seventeenth century, the Hungarian upper classes ate according to the medieval two-meal system. The meals were eaten around 9.00 or 10.00 a.m. and at 5.00 or 6.00 p.m., respectively. Both meals consisted of fresh hot dishes and had from three to twelve courses, depending on the rank of the householder. We have two complete menus extant, each covering one month, from October 1553 and January 1603. The first was prepared for the middle-class servants (noblemen and burghers) of Lord Nádasdy, while the latter comes from Lord Thurzó's household.[7] Both families were Protestant in those years, yet both observed the old Roman Catholic abstinence requirement on certain days of the week. In the Nádasdy court, Friday and Saturday were the meatless days—in the Thurzó court, it was only Friday. For both courts, we have, in addition to the bill of fare, lists of the foodstuffs, which were brought to the kitchen. Cheese did not feature in the menus at all, not even as a final course for the dinner on New Year's Day, nor, indeed, when Lord Thurzó was entertaining guests. In both courts, curds (*túró*) was exclusively the food of the days of abstinence; it was in evidence on all such days, but never on other occasions. The standard course was a hot dish of curds served on slices of bread. For this, we have the evidence of contemporary cookery books. Both fresh and dried curds could be used to prepare the dish.

The curds were mixed in with milk or sour cream, egg, and dill, or they were boiled with a small amount of water, poured on sliced bread and served (*túrós étek* 'dish of curds'). The cookery books mention that this dish was also prepared by peasants, but they made it without egg. At that time, the fashion for noodles had already reached the Nádasdy court, but it had not yet reached the kitchen of the Thurzós. In a departure from the Italian pattern, noodles were, in Hungary, associated exclusively with days of abstinence. One of the ways of serving noodles boiled in water, was with curds. Various other forms of noodles boiled in milk were also dishes eaten on days of abstinence. Milk and sour cream did find their way into the kitchen on meat-days too, but only very rarely. Butter was commonly used for cooking in the Thurzó court, though the majority of the dishes were not prepared with butter. In the kitchen of the Nádasdys, by contrast, butter was used strictly for cooking, and only on the days of abstinence. Though it is not featured in these court menus, we know of a certain 'soup of cheese' both from cookery books and from established practice. It was not eaten as a soup, however, as, prior to the late seventeenth century, the meals of the upper classes did not begin with soup; in fact, the dishes hardly even included any liquid soups. Devised by Marx Rumpolt, the Hungarian-born author of Middle Europe's representative cookery book (1581), the dish made with 'cheese soup' was intended for the Hungarian and Bohemian kings, for their banquets held on days of abstinence.[8] In Hungary, it also formed part of the provisions for the castle garrison. The dish was made from fresh, unpressed cheese, i.e. fresh curds, mixed with sour cream and butter, melted, and then poured on slices of bread to be served. Non-fresh cheese was grated into hot water, and boiled with salt and onions (*sajtlé* 'soup of cheese'). It, too, was served on slices of bread. Contemporary records of purchases contain the phrase 'I have bought curds for soup of cheese' (*túrót sajtlének*). None of the cookery books of the period planned for dishes for meat days to be made from dairy products. In the two standard meals of the day, dairy products appeared as hot dishes. The cookery books contain some further recipes using milk and curds, including pie filled with curds. Hot toast *croûtons* in cold milk were also eaten. Hot thick gruels (of millet, in the 1695 cookery book) were also eaten with cold milk. Concerning milk, curds and cheese in their natural states, a printed cookery book from the end of the seventeenth century, intended for the middle classes, states that on milk-days some of these might be served occasionally, too, and that those who like them will eat of them. But they were not regarded as courses. We know several menu-cards of guild feasts from the sixteenth and seventeenth centuries, and in only one of them—from Transylvania—is cheese also featured in the last course to round off the meal. In this case, it was probably fresh ewe-cheese.

While the adult members of the upper social strata adhered strictly to the two-meal régime, their adolescent children would be given something for breakfast early in the morning, before the pre-noon dinner. In boarding-schools, this breakfast was provided in an organized form, and it would consist of bread with curds/cheese or sometimes sausages, with a drink of wine to go with it.[9] If a nobleman was about to go on a trip, he might take with him some smoked pig fat, smoked meat, and baked chicken, with curds/cheese—which are not mentioned—obviously reserved for days of abstinence.

Craftsmen, peasants, and farm labourers, who did manual work, ate three meals during the working day, as far back as the sixteenth and seventeenth centuries. That three-meal system applied to craftsmen all the year round, and to those employed in agriculture in the summer half of the year. As regards craftsmen, we know of several cases from the seventeenth century where their additional meal, breakfast, consisted of bread and cold cheese/curds. From the sixteenth century onwards, regular bread-baking already allowed peasants to eat one bread-based cold meal during the day. However, given that even as late as the twentieth century, there were regions where, in summer, every one of the three meals of the day consisted of

Milk and milk products

fresh hot dishes, it is not easy to conjecture about a much earlier period.

In early modern times, peasants regularly made butter from the sour cream of cows' milk, and curds from sour milk. Curds were used fresh or were dried. All three were sold on the market, too. In certain regions, probably real cheese was produced as well, but it was not a dairy product widely used across the country.

In the two main meals, peasants used fresh milk with hot thick gruels and porridges. In the seventeenth century, they had learned how to make noodles and they used curds for these as well. They, too, cooked soup from curds, which they poured on bread pieces before eating them. Pie filled with curds was a festive food. Since no peasant menus have survived from early modern times, we do not know if their consumption of milk and dairy products was also heavily concentrated on the days of abstinence.

The beginnings of the custom of eating butter spread on bread can be detected in the sixteenth century in the Netherlands, Northwest Germany, and in England. It became part of popular foodways, and came to be generally accepted in a wide geographical area. In the pre-industrial civilization, the custom did not penetrate into the southern part of Middle Europe, where butter was preserved in a melted form. Bread and butter was absent from the diet in Hungary, too.

Numerous fifteenth to seventeenth century European cookery books presented their chapters according to a functional arrangement, treating in separate chapters the dishes for meat-days, as well as milk-and egg-dishes, which were allowed when there was abstinence from meat. Again, a separate chapter was devoted to dishes for days when abstinence was prescribed from all food derived from animals, with the exception of fish. These days are known as meat-days, milk-days, and fish-days respectively. The latter term seems somewhat problematic, because in numerous regions fish was a luxury item which the common people did not eat on a regular basis. On the other hand, I am trying to avoid the term 'fast', which, strictly defined, involves, in addition to abstinence, certain restrictions in the number of daily meals, as well as how far one is allowed to satisfy one's appetite. Milk-days and fish-days represented two different levels of abstinence. A late example of the cookery books compiled according to meat-days and the two levels of abstinence, is the first printed Hungarian cookery book from 1695, which, incidentally, was published by a Protestant publisher.[10]

All cookery books of this type observed the universal and regional regulations and customs of the Roman Catholic Church, which themselves altered in the course of time. As there were regional differences as well, I shall try to spotlight the Hungarian conditions in early modern times, as far as the sources allow.

In the middle of the sixteenth century a new Catholic catechism was published. Its Hungarian translation (1599) also touched on local conditions.[11] It prescribed abstinence from meat, dairy products, and eggs for Lent, for the Wednesdays, Fridays, and Saturdays during the four periods of the Ember Days (*quattuor tempora*),[12] for certain vigils, and for all Fridays of the year. Exemption from the prohibition on the consumption of dairy products and eggs was possible. All the Saturdays of the year figured as milk-days. It was noted that, in Hungary, Wednesdays were also observed as days of abstinence by many, which probably should be taken to mean that these were milk-days also. Contemporary data show that during Lent, Catholics did indeed abstain from milk, eggs, and meat. That custom was strongly criticized by Protestants as early as the sixteenth century. However, they did not criticize—in fact, as has been seen, they even adhered to—the observance of certain meatless days in the course of the week. For the time being, we might infer only from Protestant menus that, already in the sixteenth century, and in the Catholic upper classes also, dairy products and eggs were probably eaten on Fridays as well. Altogether that would mean three milk-days—Wednesday, Friday, and Saturday—in the case of Catholics—and one or two still being observed by the Protestants as well. The consumption of milk and dairy products was thus concentrated on these days. In 1611, invoking a concession granted earlier to some other countries, the Archbishop of Hungary requested, and was granted, exemption from Rome permitting the consumption of dairy products and eggs on all days when meat was forbidden. As in many other cases, the relaxation of these restrictions was certainly the recognition of what had gradually come to be common practice in Hungary. The change in the rules extended the number of milk-days: it retained the Wednesdays, Fridays, and Saturdays of every week, and added all the days of Lent to their number. By virtue of the latter development, a new period in the concentration on dairy products' consumption was initiated.

EIGHTEENTH TO THE TWENTIETH CENTURIES

From the Reformation and the Counter Reformation, the majority of the Hungarian population eventually emerged as Roman Catholic, with a significant Calvinist, and a smaller Lutheran, minority. In the twentieth century, Protestants observe only one fast-day, namely Good Friday, and no milk-days have been preserved. Even for the Catholics, the process whereby the number of abstinence days has been reduced is not known exactly. In the mid-nineteenth century, all Fridays and Saturdays, as well as the whole of Lent, were still days of abstinence from meat. Ordinary Wednesdays, as days of abstinence, no longer figured in the regulations. By around 1900, abstinence from meat came to be restricted, in the rules, to Fridays, and to only four days of Lent. We know a great deal about the popular foodways of that period and these demonstrate that the discrepancy between practice and prescriptions could also take a form where practice lags behind the (more relaxed) regulations. Until World War I, Hungarian peasants in a number of regions abstained from meat and lard throughout Lent; Lent was observed as forty milk-days, when they used butter for cooking and ate large quantities of preserved curds. To prepare for this period, they would start collecting the melted butter and salted curds well in advance.

The weekly sequence of dishes in the main meal among Catholic Hungarian peasants was recorded as far back as the middle of the nineteenth century. On Sundays, Tuesdays, and Thursdays, it consisted of meat dishes, with noodles on Mondays, Wednesdays, and Fridays, and something meatless on Saturday. That weekly sequence must have developed at a time when peasants already knew noodles, and when Wednesdays and Saturdays were still days of abstinence—i.e. sometime in the seventeenth or eighteenth centuries. The foodstuffs which provided the names for the days in the vernacular language such as *meat-days* and *noodle-days*, were not the only items in the menus. From the eighteenth century, the main meals in peasant menus, too, tended increasingly to feature soups as the introductory course, and meat with vegetables or noodles as the second course. That sort of weekly sequence of dishes did not occur in Transylvania, where noodles never acquired importance in popular culture. Here until the middle of the twentieth century, a *polenta*-like thick porridge of maize was cooked once or twice daily.

Apart from the use of butter in cooking during Lent, pig fat and lard were the standard cooking agents. Well into the eighteenth century, all the pig fat was conserved in large pieces, salted, and dried or smoked. This pig fat was eaten as such, and was also used for cooking, both in upper-class and peasant cultures. Lard i.e. fat straight from the pig which was melted down, came into use during the eighteenth century, in upper-class culture. It replaced pig fat as a cooking agent. The use of lard among peasants spread, very slowly, from the west towards the east.

Apart from Lent, butter was used, in peasant culture, only for pastries, which were associated with festivals.

In the eighteenth to the twentieth centuries, the upper and middle classes ate according to the three-meal system. Breakfast consisted of coffee with cream or milk and in the eighteenth century, dunking pieces of bread-roll or bread in coffee was acceptable in these classes. Bread with butter was not yet a standard breakfast item; this was introduced in the course of the nineteenth century, when a breakfast of white coffee, bread and butter, and jam, became one of the characteristics of these classes. Dinner, the main meal of the day, was eaten at midday. For a long time, both dinner and supper consisted of fresh warm foods. It is unclear from the data whether or not it was customary for the upper classes to finish off these meals with cheese, but certainly it was not a standard practice with the middle classes. Although the precise time of the change has yet to be clarified, it is certain that in the first half of the twentieth century it was already usual to eat a cold supper with bread, butter, cold meat and sausage, and factory-made cheese. Fresh curds mixed with sour cream could be another cold dish. The soups cooked from curds, characteristic of early modern times, had disappeared from the repertoire of dishes used for dinner. Fresh curds were used as a filling for pastry on various days, and on noodles and pancakes on noodle-days.

Modern urban industrial workers, whose number began to grow considerably from the 1880s, also had coffee with milk, for breakfast.

Peasants and agricultural labourers remained within the medieval two-meal system in winter, but from spring to autumn a three-meal system applied, with supplementary meals during periods of hard labour. Morning coffee began to be introduced from 1900, but it never became common or general in their circles. In winter, the predominant meal system was one in which the afternoon meal was the main meal, featuring, in a recurrent weekly sequence, the dishes cooked for the midday meal in summer. Breakfast, however, did not involve a daily-recurring constant dish. After pig-killing feasts, it included the pieces of meat which had not been conserved, but for the rest of the time, it consisted of various soups, thick gruels/porridges and, ever more frequently after 1830, potatoes. Fresh milk figured fairly frequently in the winter breakfast. Soups were eaten with bread or with dumplings boiled in them; an alternative to soup was warm milk eaten from a dish, with pieces of bread soaked in it. The gruel/porridge was often eaten with milk; occasionally, the gruel was boiled in milk. Boiled or baked potatoes were likewise often eaten with milk.

The summer breakfast, in the major part of the Hungarian-inhabited area, consisted of cold food eaten in the fields between 7.00 and 8.00 a.m. It consisted of bread and, whenever possible, pig fat. On days of abstinence, the pig fat was replaced with curds. In a crescent-like zone of the Carpathian Basin—a zone spreading from the southwest through the north to the southeast—the summer breakfast consisted of a cooked hot dish, eaten, in certain regions, at home in the small hours of the morning, while elsewhere it was taken out into the fields between 7.00 to 8.00 a.m. The gruel/porridge could be accompanied by milk at home, while maize porridge could be accompanied by curds when eaten in the fields of Transylvania.

Dinner, in both winter and summer, included soups which were cooked with milk. Curds could be added to noodles and the maize porridge in Transylvania. In contrast to upper-class cuisine, peasants continued to cook the soup of curds of early modern times until the twentieth century, eating it by then often as a soup proper.

In summer, fresh milk was used mainly at supper which in the early part of the twentieth century still consisted of bread pieces soaked in fresh milk eaten with a spoon. This, however, was only one of many solutions for the summer supper.

During times of hard labour on hot summer days, the afternoon snack held pride of place as regards consumption of dairy products. At mowing-, reaping- and threshing-time, peasants and their hired men regularly had an afternoon snack. Sour milk with bread was a favourite choice, making a pleasant cooling refreshment.

As the changeover to the dairy breed of cow, which happened between 1880 and 1920, coincided in time with the organization of commercial possibilities for milk, the additional milk was converted into cash, instead of being consumed by rural families.

From the 1880s, rural foodways slowly edged closer, in many respects, to those of the urban middle class. But that did not produce a spectacular change in the consumption of milk-derived products. In the 1960s, for instance, peasants were still not eating cheese.

In the eighteenth to the twentieth centuries, although small amounts of curds and milk were used in the main meals—curds in noodles and in baked pastries, and milk in certain soups—for the middle classes, peasants, urban and agricultural workers, dairy products tended to be concentrated rather in the complementary meals. As well as the overall low level of consumption of milk and dairy products in the country, the consumption of cheese and butter was particularly low. The principal forms in which dairy products were consumed were milk, sour milk and curds. Drinking milk as part of a main meal would be unthinkable in Hungary.

All in all, in Hungary, there has always been more milk in the Milky Way (the focus of numerous legends), than dairy products in the range of foodstuffs, even during the period of the many 'milky' days.

Acknowledgment:

The research for this article has been supported by OTKA (National Scientific Research Foundation), Programme No. T 4876.

NOTES AND REFERENCES

1 L. Makkai, 'A magyarországi mezőgazdaság termelési és fogyasztási struktúrája a XVII. század közepén' (The agricultural production and the structure of consumption in Hungary in the mid-seventeenth century), in P. Gunst, (ed.), *Mezőgazdaság, agrártudomány, agrártörténet, Budapest 1979, 262.*
2 K. Keleti, *Die Ernährungs-Statistisk der Bevölkerung Ungarns*, Budapest, 1887, 174, 158.
3 *Magyar Statisztikai Évkönyv* (Statistical Yearbook of Hungary), Budapest.
4 L. K. Kovács, 'Die traditionelle Milchwirtschaft bei den Ungarn', in L. Földes (ed.), *Viehwirtschaft und Hirtenkultur*, Budapest, 1969, 669, 673–674.
5 A. Paládi-Kovács, 'Milchwirtschaft in Ungarn im 18. Jahrhundert', in B. Gunda, et al. (eds.), *Ideen, Objekte und Lebensformen*, (Székesfehérvár, 1989, 198. (= *Az István Király Múzeum Közleményei* A/29).
6 See L. K. Kovács, *op. cit.*, 640–695; E. Kisbán, 'Die historische Bedeutung

des Joghurts in den Milchverarbeitungssystemen Südosteuropas', in L. Földes (ed.), *op.* cit., 1969, 517–530; J. Barabás (ed.), *Magyar Néprajzi Atlasz* (Atlas of Hungarian Folk Culture), Budapest, vol. 3. 1987, map, 179–190, vol. 6, 1989, map, 349–352.

7 See M. Belényesy, 'Egy XVI. századi főúri étrend kultúrtörténeti és néprajzi tanulságai' (The Nádasdy menus in 1553), *Néprajzi Értesítő*, vol. 40, 1958, 133–153 and B. Radvánszky, *Régi magyar szakácskönyvek* (Old Hungarian cookery books), Budapest, 1893, the Thurzó menus in 1603, 321–391.

8 M. Rumpolt, *Ein new Kochbuch*, Frankfurt am Main, 1581; Facsimile edn., Leipzig, 1976, 20.

9 P. Apor, *Metamorphosis Transylvaniae* (1736), Budapest, 1972, 85–87.

10 *Szakáts mesterségnek könyvetskéje* (The booklet of the art of cooking), Tótfalusi Publ. Kolozsvár, 1695.

11 G. Vásárhelyi (translator), *Catechismus* (By Canesius), Kolozsvár, 1599, 83–88.

12 Ember days are days of fasting and prayer especially in the Roman Catholic and Anglican Churches, on the Wednesday, Friday and Saturday after (1) the 1st Sunday in lent, (2) Whit Sunday, (3) Holy Cross Day (Sept. 14), and St. Lucy's Day (13 Dec.).

SELECT BIBLIOGRAPHY

Magyar Néprajzi Atlasz (Atlas of Hungarian Folk Culture) J. Barabás, (ed.), Budapest, I-IX, 1987–92, maps 174–190, Processing of ewe's milk; maps 346–352, Processing of cow's milk.

N. Dunăre, 'Milchprodukte im rumänischen Hirtenwesen', *Viehwirtschaft* 1969, 603–639.

L. Földes, 'Esztena und Esztena-Genossenschaft bei den Szeklern', *Viehzucht* 1961, 283–328.

L. Földes (ed.), *Viehwirtschaft und Hirtenkultur*, Budapest, 1969.

L. Földes (ed.), *Viehzucht und Hirtenleben in Ostmitteleuropa*, Budapest, 1961.

B. Gunda, 'Pflanzliche Labstoffe in den Karpaten', *Viehwirtschaft* 1969, 503–516.

B. Gunda, 'Hungarian Butter Rhymes', *International Folklore Review*, III, 1983, 34–39.

L. Keszi-Kovács, 'Die traditionelle Milchwirtschaft bei den Ungarn', *Viehwirtschaft* 1969, 640–695.

E. Kisbán, 'Die historische Bedeutung des Joghurts in den Milchverarbeitungssystemen Südosteuropas', *Viehwirtschaft* 1969, 517–530.

B. Kopczyńska-Jaworska, 'Gebrühte Schafkäse in den Karpaten', *Viehwirtschaft 1969, 531–546.*

A. Kowalska-Lewicka, 'Die Milchwirtschaft in den polnischen Karpaten', *Viehwirtschaft* 1969, 696–705

L. Kunz, 'Die traditionelle Milch- und Käsewirtschaft in Mittel- und Westmähren', *Viehwirtschaft* 1969, 706–734.

O. Lurati, 'Alpwesen und Alpbewirtschaftung im Tessin mit besonderer Berücksichtigung der genossenschaftlichen Sennerei', *Viehwirtschaft* 1969, 756–777.

V. Novak, 'Über die Milchwirtschaft bei den Völkern Jugoslawiens', *Viehwirtschaft* 1969, 574–602.

A. Paládi-Kovács, 'Milchwirtschaft in Ungarn im 18. Jahrhundert', in *Ideen, Objekte und Lebensformen*, B. Gunda *et al.*, (eds.) (= Az István Király Múzeum közleményei A/29.), 197–207.

L. Schmidt, 'Der einbeinige Melkschemel in dem österreichischen Alpenländern', *Folk-liv* XXI-XXII, 1957–58, 135–156.

I. Simonjenko, 'Almenwirtschaftliche Schafzucht der ukrainischen Bevölkerung in den Waldkarpaten im 19. und zu Beginn des 20. Jahrhunderts', *Viehzucht* 1961, 363–388.

Ch. Vakarelski, 'Milchverarbeitung und Milchprodukte bei den Bulgaren', *Viehwirtschaft* 1969, 547–573.

Gy. Viga, 'Methoden der Butterverzierung und ihre Geräte in Ungarn', *Műveltség és Hagyomány* XXII = *Ethnographica et Folkloristica Carpathica* 4 (Debrecen), 1985, 49–60.

R. Weiss, *Das Alpwesen Graubündens*, Erlenbach-Zürich, 1941.

3 Milk in the rural culture of contemporary Assyrians in the Middle East

Michael Abdalla

INTRODUCTION

From time immemorial man has been surrounded by animals. He has used them as beasts of draught or burden and as a source of food. This paper deals with the place of milk in the rural culture of modern-day Assyrians in the Middle East, but I would like to begin by considering evidence for farm animals and dairy products from the era of the Sumerians, the founders of Mesopotamian culture. Contemporary Assyrian Christians in the Middle East believe their roots date back to the times of the ancient Sumerians, and that they have had an organised existence in the Tigris and Euphrates valleys for seven millennia.[1]

Sumerian sculptures and cuneiform tablets, deciphered by the American sumerologist, S. N. Kramer, show many examples of milch animals and give information about dairy products (Fig. 3.1). A frieze from the first half of the third millennium BC from the ruins of the temple in al-Ubaid is well known to art historians; archaeologists have called it 'Dairy in al-Ubaid' (Fig. 3.2 a,b).[2] A Sumerian harp from the same period decorated with a cow's (or bull's) head, from the royal tombs in Ur, is also well known.[3] A thousand years later, animal heads mainly goats', sculpted in bituminous stone, were used to decorate the legs of great three-legged wine cups.[4]

Farm animals also feature in many Sumerian legends. In one concerning Enmerkar, the legendary builder and ruler of the city of Uruk, a priest named Shamash, talking in human language to the animals, urges the cows and sheep in the 'holy cowshed and sheep-pen' to stop giving milk to the goddess Nidaba, and the animals agree. However, two shepherds, 'brothers born of one mother', appear and throw the priest into the river, and thus thwart the priest's scheme. The priest unsuccessfully tries to hide from the shepherds' mother in the waters of the Euphrates by transforming himself successively into a sheep, a cow, and a chamois.[5]

Sumerians bred mainly sheep; a breed with a thick fatty tail sometimes reaching to the ground which often determined the commercial value of

28 *Milk and milk products*

Fig. 3.1 A votive slate, *c.* 20 cm. high, from the first half of the third century BC, which was offered by a wealthy merchant by the name of Urenlil to the Enlil temple in Nippur. The upper part shows a naked priest offering a drink (probably milk) to a bearded god sitting on a throne. In the lower part one sees goats followed by two men. One of them is carrying a pitcher (probably with milk) on his head; the other is whipping the animals with a stick (Museum of Ancient Orient, Istanbul).

the sheep, was best known. Sheep were the first domestic animals in Mesopotamia having appeared there in the sixth millennium BC.[6] The Ur tablets give the number of sheep in flocks: most frequently there were one hundred to one hundred and eighty sheep in a flock, although there were larger flocks with as many as five hundred sheep.[7] The sheep from the 'holy flock' grazed on specially designated meadows which belonged to the temple. When they were the proper weight, they were killed and the meat was eaten by the shepherds who were called 'the cattle people', or 'the sheep people', or 'the goat people'. Each sheep ate 1–1.5 *sila* of grain (0.4–0.6 litre), a few dates, and even bread, on a daily basis.[8]

Goat and cattle herds were smaller, and it was necessary, in order to maintain the quality of the herds frequently to cross the domestic animals with wild animals, e.g., cows were crossed with bisons, sheep with wild rams and goats with mountain goats.[9] The animals were divided into groups according to milk yield, fodder type and physiological condition.

Only animals which met specific criteria could be offered to the Sumerian gods: sheep had to be fed on barley, and young goats (kids) and lambs on milk.[10] In addition to listing the activities connected with breeding (for example, there was a 'sheep shearing month'), documents also mention dairy products—butter, cream, and different kinds of cheese—delivered by the shepherds (mentioned by name) to herd or flock owners.[11] Granaries of the Sumerian temple were full of different foods, including cheese.[12] Butter made from bull's milk was used in baking, and sour milk mixed with honey was used as a cough medicine,[13] and when mixed with tamarisk, it was used to relieve stomach aches and to whet one's appetite.[14]

Although detailed information on the status and use of milk and its products as food in ancient Mesopotamia is relatively scarce, nevertheless, an analysis of the literature on the history of ancient Mesopotamia reveals that milk, unlike beer for example, was neither an important food nor

Fig. 3.2 'Dairy in al-Ubaid'. A frieze, 20 cm high.

(a) A calf is put near the cow's mouth so that it gives milk more willingly.

(b) A man sitting on a stool shakes a corked container with milk to shake off the fat. The other two men strain the product to separate the butter from the whey. First half of the third century BC (The Iraqi Museum, Baghdad).

a popular drink. Generally, animal products were less important than vegetable products, and milk was no exception. It was served as cheese or sour milk for dessert,[15] and Sumerian confectioners who used butter in their products were mainly employed in temples.[16] Hence, one can suppose that the number of people who bought their products was very limited.

PROCESSING AND CONSUMPTION OF MILK TODAY

Christian Assyrians are forbidden to drink milk or eat its products for a total of one hundred and forty fasting days each year. Fresh milk is rarely drunk on other holidays: some people think it leaves a foul taste in the mouth and thus boiled milk and especially yoghurt (*qatiro*) are more popular.

Fresh milk quickly turns sour due to the relatively high temperatures in the region. It is then considered a spoiled product (*halwo hariwo*) and cannot be eaten. Therefore, after milking, it is quickly boiled. While it is boiling over the fire (*tfayo*), care is taken that the fire, made of glowing charcoal or embers, is not given to anyone else before the pot has been taken off, since it is believed that if the fire is taken to another place it becomes less efficient, and bad luck ensues for both the household and the animals. At this stage of milk processing, care is also taken to avoid demonstrating the abundance of milk so as not to evoke needless admiration and envy among others. The residue of the boiled milk in the pot (sometimes even slightly burnt) is often a bone of contention among younger children—each of them asks his mother for this delicacy.[17] It is usually given to boys and they add bread crumbs to the sediment layer and eat the mixture from the pot.

Beestings (*alwo*) is another rare delicacy for the children. After a cow has calved, the parents try to ensure that the beestings from the first milking is equitably distributed between the newly-born calf and their own children. This 'miraculous food' (it has a high nutritive value as it contains iron and vitamin A) is boiled and then given to the children. (Is this, in addition to the two to three years breast feeding of children, another explanation for the resistance of Assyrian children in the Middle East to various diseases from which European children suffer so frequently?)

Milk is one of the main ingredients for many foods. *Dasheshto* (also *zarda*) is a kind of pudding eaten by Western Assyrians. This is a thick milk soup with rice, eaten with sugar and powdered cinnamon. It is served for supper on the third (last) day of what is known as Nineveh (Lent),[13] and on the first day of Easter. A similar dish called *ryzza w khalya* is eaten by Eastern Assyrians. Milk is also used instead of water to make wheat cookies called *kilicha*, or to bake pancake-like bread.[19]

MAKING YOGHURT[20]

Yoghurt culture (*rawbe*) is added to milk,[21] which is boiled in the afternoon and cooled to a temperature of about 40°C (One dips one's finger in the milk to check its temperature). The pot is wrapped in a warm garment and left indoors to cool slowly. The slow rate at which the milk cools has a positive effect on its coagulation. In the evening the pot is taken outdoors and put in a place inaccessible to animals, especially cats which are great lovers of yoghurt, so that it can slowly set in the coolness of the night. The yoghurt has a gelatinous texture, and the more solid the texture the better the yoghurt.

Yoghurt is not usually drunk but is rather regarded as a basic ingredient for many dishes, e.g., *mparsheta* (pieces of bread soaked in yoghurt and served with sugar to the elderly who are toothless, and to babies, as a very nutritious meal). Often yoghurt is diluted with water and, with diced cucumbers, garlic, ice and salt, it is served for supper, and is eaten with bread as a cold soup. The aroma of half a date dipped in yoghurt is very pleasing; a similar aroma is produced when yoghurt mixed with date-syrup is eaten on its own, or on a slice of flat bread sometimes accompanied by sips of very sweet tea. These sweeteners—especially sugar—are used to improve the taste of yoghurt which very quickly turns sour when left in a warm place. Yoghurt is served as dessert, or, as an aid to digestion with dry and loose foods such as rice, groats or fried cereal-leguminous balls, it is served as a separate dish instead of a salad.

Yoghurt is also used as an ingredient to make *tyrkhayno*: thicker parts of *bulgur* (puffed wheat groats), yoghurt and some salt, are put into a pot and left for twenty-four hours until all the yoghurt is absorbed by the *bulgur*. Then the high viscosity pulp is stirred for some time and left for another twenty-four hours; the stirring is repeated every twenty-four hours. If the mixture, called 'dough', becomes dry to quickly, a small quantity of yoghurt is added. During the five to six days fermentation process, the 'dough' acquires a characteristic aroma and taste. Next, balls from one to four cm in size are formed from the 'dough' and dried in the sun, and this is best done when the humidity is eight to ten per cent. Low water content, low pH level, and an absence of hygroscopic features, guarantee that the product, kept in cloth sacks or open clay chambers, remains edible for two to three years. *Tyrkhayno* is served for breakfast in a diluted form which has the consistency of porridge. This dish has a sour taste, and although usually served cool, resembles a Polish sour soup called *zurek*.

Some yoghurt is used to make butter (*zebdo*) in special goatskin containers called *gawda*.

PREPARATION OF A GOATSKIN

A mature, usually male, goat is beheaded by members of the household. The skin on both hind legs is cut open crosswise and air is blown into the cuts. The air, blown in under pressure, lifts the skin off each segment of the body and separates it from the carcass. Something resembling an air cushion or balloon is formed. Next, the person cleaning the goat inserts his hand with a knife through a cut in the neck, and, without damaging the skin, removes the carcass piece by piece. The meat is consumed and the skin can be tanned in the following way: first it is immersed in a loose pulp prepared by mixing the ashes of various organic products with water which is then left in the shade for about a week. Due to the acidic reaction of the pulp the fleece can be plucked off more easily. Then the skin is washed several times and sprinkled with dried and ground pomegranate-skin (*jafte*) mixed with salt. Again it is left in the shade for a week. The tannins in the pomegranate-skin give the goatskin a specific colour. When the skin has been throughly washed and dried, it is stretched with the help of a pole (*saktho*) fixed in the ground. Another tanning technique is to coat the skin with fermented wheat-flour dough, sometimes with lentil-flour added, and leave it for a few days. The fat falls off and the hair can be plucked off more easily. Next, the skin is washed and immersed in a pulp made from boiled acorn-meal (*gamare*) for fifteen days. The four holes in the skin from the hind legs of the animal are sewn with straps of goatskin (*shubago d-mare'zo*).[22]

PRODUCTION OF BUTTER AND MILK FAT (*MESHHO*)

The production of butter takes place early in the morning when the children are asleep. It is the first chore with which rural housewives begin their day in late spring and early summer. The morning coolness helps glue the fat balls in the yoghurt.

To obtain butter from yoghurt, cold yoghurt is poured at dawn into the goatskin hung on the roofbeam with one end facing the yard. The ends of the goatskin—the remains of both hind legs—are tied to wooden grips called *lawlabe*. The *lawlabe* are joined perpendicularly to a stick (*moyuoto*). The goatskin is hung at an appropriate height from the ground. Holding the *lawlabe*, the person who is making butter shakes the yoghurt for 20 to 30 minutes until fat balls form lumps; this can be seen through the upper opening in the goatskin (Figs. 3.3, 3.4).[23] The shaking can be performed by two women standing opposite each other (Fig. 3.3). At camps the goatskin is hung on three wooden beams called *daqroye* which are beaten into the ground. The structure looks like a pyramid.[24]

Butter obtained in this way is considered a breakfast delicacy, especially when it is sprinkled with sugar. But it is not used as a spread on bread

Fig. 3.3 Making butter from yoghurt. 'Tur Abdin' region, Assyrian village, *Marbobo* (Photo: Simon Atto, 1984).

as this custom is unknown in the Middle East; instead, pieces of bread are dipped in butter and eaten. It is customary to offer some of the fresh butter to those villagers who do not have enough milk of their own. Sharing what nature gives them, and sharing their sorrows and joys, is very characteristic of the Assyrians.

Butter obtained from yoghurt is mainly used to make non-perishable milk-fat which is added to dishes served on holidays throughout the year. The butter is salted and put into a container. When the container is full (about 20 kg), the butter is taken out, put into a thick-walled, tin-plated tub, and then boiled. When boiled the milk falls into layers—pure fat forms on the top and whey and sediments collect underneath. When the milk has cooled, the fat is separated and boiled again, this time with a piece of bread. The boiling is stopped when the bread becomes crisp and light brown in colour. The fat is stored in ceramic, metal, or leather containers. Before it is served, it is warmed and then poured over the dish. An average Assyrian family eats about 60 kg of this very tasty fat a year.[25]

Rhythmic songs about the traditional ceremony of butter making have been written. Of two such songs in my own collection, one is entitled *gudi mayyanne b dranani* (I am shaking the goatskin with the strength of my arms), and is performed by a woman, while the other, performed by a

34 *Milk and milk products*

man, describes the path along which a man's sweetheart can go, and on which she finds a goatskin.

WHEY DISHES AND PRESERVES

When the butter has been separated, whey (*dawghe, dawwe*) remains in the goatskin, and this is drunk, often instead of water. It is also given to the needy. When there is a lot of it, it is given to farm animals, or

Fig. 3.4 Making butter from yoghurt. 'Tur Abdin' region (central-southern Turkey), Assyrian village, *Mzizah* (Photo: Simon Atto, 1984).

poured into containers and sold in nearby cities. It is mainly bought by those who have recently migrated from villages to towns, who drink it with ice, or puts pieces of stale bread into it and then eat it. The Assyrian menu contains a few dishes with whey, and they are considered wholesome meals. *Labaniye* is a Western Assyrian summer dish served cool for supper. It is made from hulled wheat boiled in salted whey until it is soft; stirring prevents the whey from turning sour. A similar dish, though richer, is known among Eastern Assyrians as *bushala d dawwe* and consists of wheat or corn, egg-white, diced beef or lamb, boiled in salted whey. For *gierdo*, another dish, whey is boiled with an egg and salt, then rinsed rice is added and stewed until soft. The pot is taken off the oven, covered with a towel for half an hour, and then the dish is served with date-syrup or melted butter.

Whey is used to make wheat-dough which is used to bake *gyrdayye* bread. It is also used to make a dehydrated mass called *shyrtan* or *qashka*, or a kind of fermented curd called *jajiq*. The *shyrtan* is made in the following way: raw whey is poured into sacks which are hung in the sun until the liquid has evaporated. The remaining thick mass in the sack is then salted and formed into balls, the size of a hen's egg, and these are then dried in the sun. During winter the *shyrtan* is used to prepare *gabula* (Western Assyria) or *dokhwa* (Eastern Assyria), which is eaten for lunch, as follows: the hard balls are crushed in water against the coarse walls of a hand-made wooden or baked clay basin and the small lumps that remain are given to the children. Next, hulled wheat grain (*garso*) is boiled until soft, then put into a pot and hollowed out so that the liquid left after the dehydration of *shyrtan* which is poured into it, does not spill out (the edges of the wheat mass are higher and thicker.). Heated milk-fat is spread over the liquid. When *dokhwa* is made, hulled barley seeds are used instead of wheat.

Jajiq is made in the following way: whey is boiled, poured into sacks, and then kept in the shade for four to five days; to speed up the draining of the liquid the sack can be pressed with stones. Then the remaining solids are salted and mixed with sliced leaves of green garlic. The mass is beaten in a clay container and its opening is then sealed with red clay. The container is placed upside down in a pit inside a room or in the ground outside so that moisture can be absorbed. *Jajiq* is ready for use after one to two months and it is mainly served for breakfast. It has a loose texture without visible fungi and it tastes like the Polish cheese *rokpol*.

MAKING CHEESE

When I think about cheese-making, I recall my childhood and a legend which my mother told me about a certain traveller who, supposedly, was born in Mesopotamia. He filled a flask made of a dried sheep's stomach with milk, and set out on a long journey. When, exhausted, he stopped to rest and drink some of the milk, he discovered that there was no milk left in the flask but only a white lump—cheese! He tasted it and started to enjoy it. (I was very pleased when I read this legend in a book published in Europe but unfortunately, I did not have the chance to tell my illiterate mother this.)

In Assyrian lands in the Middle East, spring brings happiness to nature and people, but not to every domestic animal. A sheep, goat or cow which has given birth must reconcile itself to the loss of its offspring, especially if it is male (*qarquro, safuro, arnuwo*—male gender; *qarqurto, safurto, goleke*—female gender). The farmer lets the young animal suck its mother's milk for a day or two, but then it is killed—away from its mother's eyes. No food other than the beestings is allowed into the animal's stomach. In fact the young animal is killed for its stomach or, more precisely, for the curd which has formed in it from the coagulation of milk by acid or rennet (stomach enzyme).

The animal bereaved of its young, (especially in the case of a cow), often becomes disobedient or restive, and refuses to give milk. However, it can be cheated by the farmer, who stuffs the skin of the slaughtered animal with hay and rags and places it close to the mother. Frequently, the animal is tricked; it licks the dummy's skin and becomes less resistant during milking.

After the young animal has been killed, if it is found that there is not a sufficient amount of curd in its stomach, more milk is poured into the stomach and left to curdle. This stomach content is used as a cheese dressing called *fyrshyk*. It can be used immediately it has been made, or it can be preserved as follows: some salt is added to the *fyrshyk*, then it is put into a sack made from a white cloth, or left in the stomach and dried in the sun. This dressing (*darmono*) is immersed in milk heated to a temperature of 40°C, it is then stirred and left for about twenty minutes.[26] The whey that forms after curdling is poured away or fed to the animals, or it is often used instead of water to make a dough for *kilichaa* cookies.[27] The cheese is put into 20 x 20 cm white sacks and pressed with stones so that the liquid can flow away. The little bars of cheese that form are cut into slices the size of a matchbox, or they are heated until soft and then cut and pressed by hand; the remaining whey drains away. Cheese pieces are then formed by hand. The cheese can also be crumbled and poured into a frying-pan containing very little water, which is placed on the fire for the heat to melt the cheese. This is then formed into thick ribbons which are plaited or looped. The cheese obtained as a result of this heating technique has a spongy and solid texture and is very pleasant to eat.

All types of Assyrian cheese are white in colour. They are stored, often with fennel flowers (*Nigella sative*), immersed in saline water in glass or metal containers—the salt concentration is usually determined by floating a hen's egg in the liquid—and the tops should protrude only about one cm above the water level.[28] The cheese is eaten with bread and tea, or with water melons as a dietary dish.

MOTHER'S MILK—SOMETHING BETTER?

Much more importance is attached to mother's milk in the folk culture of the Assyrians than to any other food. The metaphysical symbols of mother's milk are present at different periods and situations in one's life. One often hears children swear an oath 'on mother's milk' to convince others of the truthfulness of what they have said. Very often a doubtful child demands that his/her friend swears an oath 'on mother's milk', which is a guarantee that the information or tales are true; it is considered a mortal sin to abuse this oath.

A baby is breast-fed for a fairly long period of time, from one to three years. It sometimes happens that one woman breast-feeds two babies at the same time, only one of whom is her own. There would be nothing strange about this if it were not for the belief that it is not advisable for a boy and a girl fed on the same breast to marry in the future, despite the fact that they are completely unrelated. This means that mother's milk shared by unrelated babies is believed to make them siblings.

An honest, truthful, loyal and well brought up man is said to have 'sucked true, uncontaminated, legal milk'. What is meant here is that a man's conduct reflects the permanence of his mother's love which was given to him as a baby during breast-feeding, and loyalty acquired during the first months of life. Irresponsible, treacherous people, are said to have 'sucked false, contaminated, illegal milk'. When children have been extremely naughty, some mothers say to them: 'May my milk be illegal to you!' This is understood as a curse. Perhaps there is nothing more forceful, or at the same time more painful, for a child's psyche than his/her mother's disapproval, expressed so firmly. When a child hears this, he/she usually cries, apologises and is submissive until the mother forgives him/her.

CONCLUSION

This description of milk and its role in the folk culture of contemporary Assyrians in the Middle East, given against a historical background, does not by any means exhaust the topic. It pertains only to Western Assyrians from Tur Abdin whose experience with milk is, in my opinion, very broad. During a short vacation in the summer of 1991 in Sweden (inhabited by about forty thousand Assyrian immigrants), I heard many stories about milk. In almost every Assyrian home in Sweden I witnessed the making of yoghurt and the preparation of *labaniye*. Some families look forward to price discounts on milk so that they can buy about thirty litres of it and make cheese from it. A neighbourly feeling of co-operation exists when word of price discounts spreads, and the address of the store is passed around. At such a time assistance in the transportation of milk, and the preparation of cheese, is offered. Assyrian families use ready-made marinade in tablet form. Some milk dishes are even served in Assyrian restaurants in Europe. In the Orient Restaurant in Enchede, Holland, I came across *jajiq* and *gweto*, served in various ways.

I think that Assyrian folklore related to milk will live on for many years, even in Europe where there are no museums of Assyrian folklore. The technology of milk processing practised in villages, and the specific customs of the people, have been recorded on tapes and in theatre scripts. However, it is worrying that ancient Assyrian villages in Tur Abdin are becoming deserted, and that children of former farmers, born outside their home countries, will have but few chances in the future to touch authentic *gawdo*, or to see how *shyrtan* is dried in the sun, or how a *jajiq* moulds in the pit of a habitable room.

NOTES AND REFERENCES

1 1992 AD is equivalent to the year 6742 in the Assyrian calendar. The Assyrian year begins on April 1st.
2 H.W.F. Saggs, a well known British assyriologist, in *The Greatness that was Babylon*, N. York, 1962. (Polish edn., 'Wielkość i upadek Babilonii, Warszawa 1973, 159), makes the following comment on the scene: 'Sumerians proved much better observers of animal behaviour than the Scythians two thousand years later. According to Herodotus, slaves blinded by the Scythians took bone blow-pipes and put them into the mare's genitals. One of them blew air while the other milked the mare. They did this, they claimed, for the following reason: the mare's veins swell when air is blown inside the genitals and its udder expands'. (See also Herodot, *Dzieje* [History], Warszawa 1959), vol. 1, 273). All page references in this and following notes are from the Polish editions of the cited texts.
3 1m long harps can be seen in the Iraqi Museum, Baghdad, and at the British Museum, London. (Cf. M. Bielicki, *Zapomniany świat Sumerów*, 401, illust. 66 (The Forgotten World of the Sumerians); Saggs. *op. cit.*, 7, illusts. 6, 13).
4 A three-legged wine cup can be seen in Louvre, Paris, (Saggs, *op. cit.*, 8, illust. 26).
5 Bielicki, *op. cit.*, 291.
6 J. Zabłocka, *Historia Bliskiego Wschodu w starożytnności*, Wrocław, 1982, 15 ('History of the Ancient Middle East').
7 Bielicki, *op. cit.*, 290.
8 *Ibid.*, 284, 290.
9 *Ibid.*, 293. A fairy tale in which a lion is cheated by a clever goat serves as an illustration of the care taken of the livestock (*Ibid.*, 296). Minutes from a '180 sheep court case' is another illustration, (*Ibid.*, 359). In the second tablet of an Epos on Gilgamesh, Enkidu who 'only knew grass on the pasture and sucked udders' goes to the city of Uruk to fight against Gilgamesh, had 'stronger bones and thicker hair because he sucked the milk of wild animals'. (A. Dobrovits, L. Kakosy, G. Komoroczy (eds.), *Babel Tornya*, Budapest, 1964; (Pol. edn. 'Wieza Babel', Warszawa, 1975, 81, 2nd ed.)
10 A. Mierzejewski, *Zapisane na glinie* ('Written on Clay'), Warszawa 1979, 143.
11 Bielicki, *op. cit.*, 291.
12 Saggs, *op. cit.*, 416.
13 Saggs, *ibid.*
14 Mierzejewski, *op. cit.*, 257–8.
15 G. Contenau, *La vie Quotidienne à Bablone et en Assyrie* ('Everyday Life in Bablyon

and Assyria'), Paris, 1950; (Polish edn. 'Życie codzienne w Babilonii i Asyrii', Warszawa 1963, 68).
16 *Ibid.*, 88.
17 Slightly burnt or roasted products are very popular among children. It is probably the product's nut-like taste which is formed when protein residue is joined with fatty acid that attracts children to it. When boiled milk is left in the sun it takes on a nut-like taste and smell. The same taste and smell is then acquired by the yoghurt. This practice is very typical of the Bedouins. I can still recall my mother who encouraged me to eat burnt bread because, she said, it would give me *hemre* (the beads) which it supposedly contains. (The glass beads game was very popular among Assyrian youth in the 1960s.)
18 Cf. M. Abdalla, 'The Way the Contemporary Assyrians in the Middle East take Food during Fasts and Church Holidays', in H. Walker (ed.), *'Feasting and Fasting', Proceedings of the Oxford Symposium on Food and Cookery 1990*, London 1990, 15–26.
19 This kind of bread called *chubz muda'am* has been known for a few years only. It is baked in small quantities in special bakeries, and because it is fairly expensive, it is supposedly made for the authorities.
20 Yoghurt is very popular in the Middle East. In Syria, Lebanon, Jordan, Israel and Egypt it is known as *kishk*, in Iraq as *kushuk*, and in Turkey as *tarhana*. There is extensive literature about *kishk*. The following authors have written about it and its nutritive values: M.M.A. Al-Shabibi, A.M. Siddiqi, *et al.*, 'Studies on the Sumach (Rhus hirta) of Iraq', *Canadian Institute of Food Science and Technology Journal*, vol. 15/1, 1982, 65–7. Sumach is used as a seasoning for *kishk*; Y.S. Hafez and A. S. Hamada, 'Laboratory Preparation of a New Soy-Based Kishk', *Journal of Food Science*, vol. 49, 1984, 197–8; A. S. Hamada and M.L. Fields, 'Preliminary Evaluation of a New Type of Kishk Made from Whey', *Journal of Food Science*, vol. 47, 1982, 1140–2; M. Abdalla, 'Kiszk—sfermentowany produkt pszenny Bliskiego Wschodu' (Kishk—a fermented wheat product of the Middle East), *Przegląd Gastronomiczny*, 1986, no. 3, 26–8. *Kishk* was also the topic of a master's thesis by Barbara Nadolna, Institute of Vegetable Food Technology, Agricultural Academy in Poznań, in 1986. The author suggested peas as a substitute for bulgur.
21 Better quality and less perishable butter is obtained from sheep's milk, rather than from goat's or cow's milk. The milk quality of black and white cows is better than that of brown cows, and the whitest yoghurt is obtained from goat's milk.
22 An identical type of goatskin used to store loose products is called *mzido*, or *mzitho* and one used to store water is called *qarbo*.
23 On average, 1 kg of butter is obtained from 10 kg of yoghurt. When dawn breaks, many swallows (*slawnoto*) appear around farmsteads. It is believed that they bring luck to the household, and increase the output of butter. Children are told that their twittering is a chorus of praise for the sunbeam and the morning prayer to Christ. Swallows often make their nests inside the households under the wooden and straw ceilings. Children are warned by their parents not to do any harm to the swallows as this is considered a mortal sin. Similar warnings were voiced as early as the thirteenth century by Bar Ebraya, an Assyrian philosopher, in his *Hudoye meṭūl qonunenune 'ionoye w-nomuse 'olmonoye* An Introduction to Church Canons and Human Laws, Bar Ebraya Verlag, Holland, 1986, 275–281.
24 In villages in north-eastern Syria inhabited by Arab Bedouins, the goatskin with yoghurt is put on a mattress placed on the ground and is shaken by women who sit rather than stand.
25 There is a popular custom practised by Eastern Assyrians on a wedding day. Before the bride enters the bridegroom's house she stops for a while in front

40   Milk and milk products

of the door, dips her finger in the *meshto*, and makes the sign of the cross on the door frame. This is said to bring happiness and wealth to the newlyweds.

26 In 1991 I was a witness to cheese making when I stayed with some Assyrians living in Holland. There the liquid dressing is bought from a wholesale warehouse called *Togo* (the Crown) run by an Assyrian from Turkey. The dressing is poured into a vessel. Milk direct from the farm is added to the vessel.

27 *Kilicha* cookies made with whey are especially popular among Assyrians from Urhoy, Edessa (today Urfa in Turkey), who nowadays live in a housing estate in Aleppo, Syria.

28 A marinade for preserving vegetables is prepared in the same way.

4 Milk products in the everyday diet of Scotland

Alexander Fenton

In previous papers, I have spoken of milk and milk products of traditional types. I looked in some detail at the Northern Isles, Shetland and Orkney, and also at the preservation of milk products in the form of butter and cheese, and of other products derived secondarily from buttermilk.[1] The present paper looks at the broad relationship between milk products of traditional type—i.e. those associated with the domestic economy in areas of subsistence or small-scale farming—and the effect on them of commercial products based on the development of dairying. Such development came to a great extent in response to industrialisation and urban growth that created a demand from ever increasing numbers of non-producers from the late eighteenth century onwards.

A great deal of information is available; it can only be handled in a simplified way here. Two preliminary points to be made are:

1   The exploitation by the human race of the milk of livestock, especially cows, as a major dietary element, is in itself a remarkable achievement that makes it a worthy subject for ethnological study.
2   The drinking of fresh milk by itself, cold or warmed, or sometimes mixed with water[2] was probably—leaving aside its role in breast-feeding—the least important aspect of the uses of milk. Beyond this, all subsequent forms of milk have a conservation-inspired basis, that spread its availability over a longer period of time, and also adapted it for distribution or trading over wider areas. Modern technological processes have greatly extended both of these factors.

The model in Figure 2.1 reflects the traditional situation in Scotland. It does not take into account techniques known elsewhere, such as the souring that produced yoghurt, though the souring and fermenting of secondary milk products were not unknown.

The model[3] reveals a double pattern: immediate consumption for the drinking of fresh milk, and delayed consumption for milk and butter, and for the drinks and products further derived from buttermilk and whey. The range of stages at which these by-products could be drunk

Milk and milk products

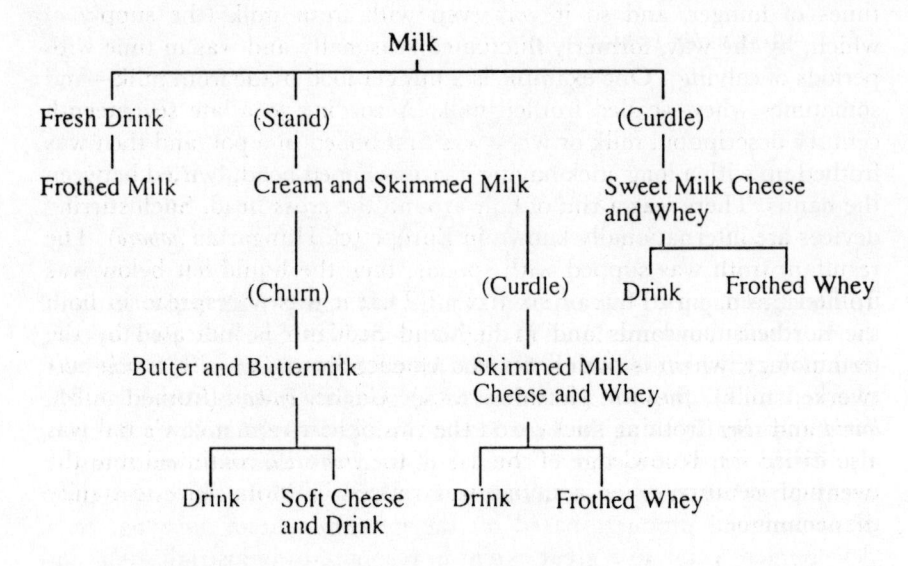

```
                            Milk
          ┌──────────────────┼──────────────────┐
      Fresh Drink         (Stand)            (Curdle)
          │                  │                  │
      Frothed Milk   Cream and Skimmed Milk   Sweet Milk Cheese
                            │                  and Whey
                            │                    │
                         (Churn)    (Curdle)  ┌──┴──┐
                            │          │    Drink  Frothed Whey
                    Butter and Buttermilk
                            │      Skimmed Milk
                            │      Cheese and Whey
                      ┌─────┴─────┐   ┌─────┴─────┐
                    Drink   Soft Cheese  Drink  Frothed Whey
                            and Drink
```

Fig. 4.1 Model of traditional milk products in Scotland.

is significant. They could be drunk immediately or could be allowed to ferment. Either way, there is little doubt about their ancient lineage and widespread nature, not only as liquid refreshment for human beings, but also as a food for livestock, in particular pigs and poultry. Going even further in the chain of forms of utilisation, buttermilk was also used for baking (e.g., of flour scones), and boiled whey might serve instead of water for making the oatmeal dish *brose*, or was boiled with oatmeal in a pot to make porridge.[4] The surviving or documentary traces of such usages point a sure finger at the extent to which the by-products—buttermilk, whey and their derivatives—were extensively integrated into the everyday diet of our ancestors from early times down to the twentieth century. The primary products, however—butter and cheese—had other uses as well, notably as a source of barter and trade and for paying at least part of the tenant's rent in kind. Such payment, called *cane* from at least the twelfth century in Scotland, could refer to all kinds of produce from the stock and soil, but in the predominantly pastoral areas of south-west and west Scotland, cheese was a major element. This was true to such an extent that still in the late nineteenth century the name *kainer* was applied in the south-west to a dairyman who paid his rent in cheese.[5] And further back, in the seventeenth century, the highland areas were said to have supplied the lowlands with cheese at times when cereals were scarce.[6]

It should not be thought that the highlands always had plenty. It is in the nature of traditional food practices that they are adapted during times of hunger, and so it was even with fresh milk (the supply of which, by the way, formerly fluctuated seasonally and was in tune with periods of calving). One example is a hunger food made from milk—and sometimes whey—called frothed milk. According to a late seventeenth century description, milk or whey was first boiled in a pot, and then was frothed up with a long stick having a cross-shaped head, twirled between the hands. There was a rim of hair around the cross-head. Such stirring devices are internationally known in Europe (cf. Hungarian *habaró*). The resultant froth was supped with spoons; then the liquid left below was frothed again, up to five or six times.[7] That it was widespread in both the northeast lowlands and in highland Scotland is indicated by the terminology, which is both Scots and Gaelic; Scots: *froh-milk, wrocht-milk* (worked milk), *fro-stick*, (frothing stick); Gaelic: *omhan* (frothed milk), *loinid* and *ròn*, (frothing stick), and the rim of hair from a cow's tail was also called *ròn*. Knowledge of the use of the *froh-stick* continued into the twentieth century, when a rhyme was related in Morayshire to signify the technique:

> Not too high, not too low,
> not too fast and not too slow.[8]

The practice was clearly well integrated into the popular community culture.

I am giving this specific example because it emphasises the importance of making maximum use of milk resources at a period when, and in areas where, milk-producing stock for the most part served *individual families*. The contrast could hardly be greater with the increasing commercialisation of milk production from the late eighteenth century onwards, in the service of urban groupings, miners and other industrial workers.

Nevertheless, there remain forms of regional continuity. I have mentioned the emphasis on butter and cheese as part of the rent in kind in west and south-west Scotland. It was in these areas, but especially the south-west, that moves towards what can be called dairying first began.

John Skene of Hallyards in Midlothian, near Edinburgh, writing ante 1666, gives an early clue. Speaking of the relative merits of cow's milk and sheep's milk, he said the former was best for butter and the latter for cheese. And he added: 'They use in Cuninghame to make cheiss of kys milk, but it is not good'.[9] Until the late eighteenth century, sheep's milk was a main basis for cheese, but a century earlier, Cunninghame in Ayrshire had gained some fame, or perhaps notoriety, for its cow's milk cheese. This is exactly the area that gave birth to the dairying industry in Scotland.

This development in south-west Scotland is a story of much ethnological interest. It ties in with an area more suitable for grazing than for arable, lying in the neighbourhood of growing centres of urbanisation and industrialisation. It demonstrates a series of links between technology (milking machinery, refrigeration, transport), the size of dairy herds, the size of farms and of the work force required, estate and tenant farming organisation, government legislation and organisation (including concepts of welfare), cattle breeds, and much else, all of which deserve more extensive study.

The historical pattern of development opens with John Skene's comment on Cunninghame cheese ante 1666. By the end of the following century, dairying was the dominant activity there and in neighbouring districts, and the making of Dunlop cheese, based on cow's milk, was spreading. There was a growing demand for butter and cheese in Glasgow and Paisley markets, and supplies even reach the remoter Edinburgh market.

A double pattern of supply can be distinguished:

1. Dairy farmers living close to urban industrial districts, including mining areas, could sell fresh milk or (in line with the relatively limited nature of the use of fresh milk) more often buttermilk and butter. The use of buttermilk, known as *soor dook*, became a widespread urban phenomenon; every urban household had its *soor dook* cans to collect supplies from the buttermilk churns brought from the countryside around in horse-drawn *soor dook* carts, well into the period of living memory. The whey was used in baking, and as a drink, as also happened in Dublin.[10] The term *soor dook* goes back to the 1820s.

2. Remoter farmers concentrated more on the conversion of liquid milk into cream cheese or skimmed milk cheese. It can be calculated that the average number of stones of cheese per cow in Cunninghame and Kyle was between 21–23 stones (230–250 kg) per annum (Ayrshire stones were of 24 lbs each). Some of this paid the rent. As an alternative in remoter districts, e.g., Wigtown, much of the milk was fed to calves, and butter and cheese production was more limited; the calves were sold to form parts of dairy and cattle herds elsewhere.

Nothing stands still, of course, and new economic factors led to nineteenth century change. The distinction between the two sides of the pattern quickly began to be eroded with the spread of the railway network from the mid-nineteenth century. Increased profitability had a knock-on effect on breed improvements. The Ayrshire breed, prolific in milk, became well established, an Ayrshire Cattle Herd Book being begun in 1877. From the 1850s Cheddar cheese began to replace Dunlop cheese, largely through the efforts of the Ayrshire Agricultural Society (established 1835), that set up direct links with Somerset in the south of England. With the encouragement of the Highland and Agricultural Society of Scotland, instruction in cheese making, and to a lesser extent in butter making, was being provided widely in the South West. A Dairy School was established at Kilmarnock in 1899, and a commercial creamery was set up in the 1880s at Dunragit, capable of handling the milk of up to two thousand cows, and feeding the by-product, whey, to a stock of up to five hundred pigs.

Transport improvements and education went hand in hand. But the scale of operations remained relatively small as long as the main dependence was on hand milking. The basis of the dairy industry was family-sized farms: in the 1870s, two-thirds of the south-west holdings were under one hundred acres. They were effective as small units because they had to stand the pressures of hand-milking, seven days a week, three hundred and sixty-five days a year; a milker, usually female, could milk an average of ten cows in one and a half hours. The units were run by families with an unmarried servant, the indoor dairy work being done by wives and daughters. The average labour force consisted of one ploughman, an odd-job man who also assisted in the byre and delivered milk, a seasonal harvester, and four women workers.

It is a remarkable fact that the pressures of coping on the basis of family labour alone led to the adoption of a system known as *bowing* (from *bow*, a herd of cattle) which lasted sporadically until after the Second World War. This meant that the farmer provided land, grazing, feeding stuffs and stock to a *bower*, who was paid for running the farm and the herd in money or in kind. This practice became common after the Napoleonic Wars on small family farms with little capital of their own, and perhaps insufficient manpower.

It faded after the 1940s as farmers took dairies under their own control, employing a full-time dairyman or dairymaid for cash wages, sometimes with sharing of the proceeds above a certain level of return. The setting up of commercial creameries had reduced the need for *bowers* in any case, as did the adoption of milking machinery.[11]

In livestock farming, milking machinery—along with shearing machines for sheep and equipment for the broiler industry—was the main form of mechanisation. Experiments had been going on in Germany, Denmark and Sweden from the 1840s, and an American machine was exhibited in 1862. But it is no surprise that a prominent role was also played in south-west Scotland. In 1891 Stewart Nicholson of Bombie, Kirkcudbrightshire, invented a teat-cup based on a cow's horn with india-rubber pads for milking; in 1899, William Murchland of Kilmarnock patented a milking machine. Both involved continuous suction, however, and it was not until 1895 that Dr Alexander Shields of Glasgow incorporated a pulsator that reproduced the sucking action of the calf. An improved version, the

46   Milk and milk products

Kennedy and Lawrence Universal Cow Milker, tried out in 1900, proved to be amongst the world's first really practical milking machines.[12]

In this way, the south-west made its mark on the dairy industry not only within Britain but also farther afield. By 1967, there were in the area 36,329 milking machines, or 61 per cent of the Scottish total. The area influenced Scotland in the technology and methodology of dairying, and in the provision of dairy herd replacements to other areas, notably the Northern Isles and North east Scotland, which have 'flying herds', i.e. they are not as a rule big enough to provide their own replacements.

Nowadays dairying, based on Friesian and Ayrshire cattle, is concentrated in the south-west, though Fife/West Lothian and the Aberdeen hinterland are also important dairying areas. District Milk Marketing Boards have been set up. Collection, transport, distribution and manufacturing of milk have been rationalised in connection with creameries.

There are a few aspects of the Scottish industry that make it different from the rest of Britain. It is less important overall, because there is a smaller area of land suited to dairy farming and a smaller population. It is self-sufficient, and after the three Milk Marketing Boards were created in the 1930s no Scottish milk was sold across the border. Scots drink less milk per head than in England and Wales—the figure was 89 per cent in 1965/6—and a large proportion of Scottish milk is manufactured in the creameries into butter and cheese. Finally, it is worth noting that milk is, or was, the second most important source of income overall, amounting to 28 per cent of gross farm income in 1965, and 20 per cent in 1970. The peak period for dairy cows was 1950, when there was a stock of 371,424. The subsequent decline continues.[13]

For the moment I do not want to go into more detail about the modern situation. My purpose has been rather to consider the interface between the old world of traditional practices, and the modern commercial world which itself has developed in clear phases. It is remarkable that in spite of changes that have been accumulating from the late 1800s, knowledge, and even practice, of older traditional ways has still survived as a fascinating demonstration of persistence. Our new world, increasingly EC-conditioned, still does not forget the old, but the older forms of direct contact between producer and customer, which ran into the earlier stages of the commercialisation of the industry, have completely gone.[14]

REFERENCES

1   A. Fenton, 'Traditional Elements in the Diet of the Northern Isles of Scotland', in *Ethnologische Nahrungsforschung. Ethnological Food Research*, Helsinki, 1975, 64–78; A. Fenton, 'Sequences in the Conservation of Milk products in Scotland', in A. Riddervold and A. Ropeid (eds.), *Food Conservation. Ethnological Studies*, London, 1988, 195–199.
2   Cf. *Kulturhistorisk Leksikon for nordisk middelalder* XI, 1966, sv. *Mjolkhushållning*.
3   For a more generalised and wider ranging model, see M. Ryder, 'Milk Products',

in M. Jones (ed.), *Integrating the Subsistence Economy*, BAR International Series 1983, 243.
4 Cf. *Scottish National Dictionary* IV, 1956, s.v. *Fey*, n.[2] (nineteenth-twentieth century sources).
5 Examples in *Dictionary of the Older Scottish Tongue* I, 1937, s.v. *Cane*; *Scottish National Dictionary* V, 1960, s.v. *Kane*.
6 C. S. Terry (ed.), Sir Thomas Craig, *De Unione Regnorum Britanniae Tractatus* (Scottish History Society), 1909, 447.
7 M. Martin, *A Description of the Western Islands of Scotland*, London, 1716, (reprint Edinburgh 1981), 202–203.
8 *People's Journal* (10 Nov. 1951), quoted in *The Scottish National Dictionary* IV, 1956 s.v. *Fro*.
9 A. Fenton, 'Skene of Hallyard's Manuscript of Husbandrie', in *Agricultural History Review*, XI no. 11, 1963, 69.
10 P. Doyle and L. P. F. Smith, *Milk to Market. A History of Dublin Milk Supply* (Leinster Milk Producers Association), Dublin, 1989, 5.
11 R. H. Campbell, *Owners and Occupiers. Changes in Rural Society in South-West Scotland before 1914*, Aberdeen, 1991, 60–93.
12 A. Fenton, *Scottish Country Life*, Edinburgh, 1976, 156; see also Doyle and Smith, *op. cit.*, 161–3.
13 J. T. Coppock, *An Agricultural Atlas of Scotland*, Edinburgh, 1976, 30, 32, 34, 125–7, 132, 231; J. A. Symon, *Scottish Farming Past and Present*, Edinburgh and London, 1959, 429–431.
14 For an example from another subject area, see W. E. Roberts, 'Folk Crafts', in R. M. Dorson (ed.), *Folklore and Folklife. An Introduction*, Chicago and London, 1972, 235–6.

SELECT BIBLIOGRAPHY

R.H. Campbell, *Owners and Occupiers. Changes in the Rural Society of South-West Scotland before 1914*, Aberdeen, 1991 (much detail on the dairy industry of south west Scotland).
V. Cheke, *The Story of Cheese-Making in Britain*, London, 1959.
A. Fenton, 'Traditional elements in the Diet of the Northern Isles of Scotland', in *Ethnologische Nahrungsforschung. Ethnological Food Research*, Helsinki, 1975, 64–78.
A. Fenton, *Scottish Country Life*, Edinburgh, 1977, Chapter 8 (Milk, Butter and Cheese).
A. Fenton, *The Northern Isles: Orkney and Shetland*, Edinburgh, 1978, Chapter 53 (Cattle and Milk Products).
A. Fenton, 'Sequences in the Conservation of Milk Products in Scotland', in A. Riddervold and A. Ropeid (eds.), *Food Conservation. Ethnological Studies*, London, 1988, 195–199.
W. Harley, *The Harleian Dairy System*, London, 1829.
Sir T. Hunter, *Statutes and Regulations affecting dairies and milk supply in Edinburgh*, Edinburgh, 1898.
J. McMaster, 'Scotch Cheese Making', *Transactions of the Highland and Agricultural Society of Scotland* XVII, 1985, 214–218.
I. Macleod et al., *The Scots Thesaurus*, Aberdeen, 1990, 212–213 (terminology of dairy products).
G. Michie and A. Fenton, 'Cheese Presses in Angus', in *Scottish Studies* 7 no. 1, 1963, 47–56.
R. Urquhart, *History of the Scottish Milk Marketing Board*, Paisley, 1979.

5 Milk dishes and World War I: the revival of traditional recipes in Germany?

Barbara Krug-Richter

The Allies' blockade which had more or less cut off the German Reich from foreign supplies since 1914, had fatal consequences particularly for the food sector. The enforced national autarky led to severe supply and distribution problems between urban and rural regions, and between food surplus and food deficit areas. As early as 1916, the population of many German towns was on the verge of starvation, and the temporary famines of 1917 and 1918 together with the years of undernourishment and malnutrition, had fatal effects on the health of the majority of at least the urban population.[1]

Since 1915, there had been deficiencies in the meat, fat and wheat supply. National and communal authorities believed that this could only be compensated for if they succeeded in 'declaring a state of war' in the households. The 'war kitchen' became the 'battlefield of the housewives', the so-called 'home front'. The women who stayed behind were cast in important roles: they were supposed to contribute to the endurance, and thus the victory, of the German people, by economical housekeeping, but also by skilled preparation of meals which were well-adapted to the specific socio-cultural needs of the time.[2]

Taking the examples of milk and grain dishes and their places in the war-time menu, this paper attempts to analyse the changing nutritional conditions and their influence on food habits during the war period on three different levels. First of all the numerous normative suggestions to the solution of the problems of food supply contained in a lot of official and unofficial pamphlets, especially cookery books, will be considered. Secondly, these normative proposals are contrasted with the nutritional reality and actual consumer behaviour which is observable in contemporary household budgets. Finally, individual solutions to the lack of basic foodstuffs as they are reflected in autobiographical testimonies are compared with the normative propositions and consumption data. The aim of this specific approach is not only to get as complete a picture as possible of the changing food habits concerning a special kind of dish, but also to have a look at the relationship between normative suggestions and their realisation.

THE NORMATIVE WAY TO OVERCOME THE CRISIS: COOKERY BOOKS

In addition to general food propaganda in newspapers and leaflets, a new genre of functional literature boomed: war cookery books.[3] Between 1914 and 1918 alone, at least two hundred specific war cookery books appeared, many of them issuing from different associations. The booklets were written by many well-known authors of cookery books, by headmistresses of domestic science schools, or by women who fully devoted themselves to the national cause. In their prefaces and general instructions, they supported public and communal propaganda, which primarily aimed at changing the food habits of the population in order to overcome the difficulties in food supplies.

The majority of the booklets were published in 1915, the second year of the war.[4] Their number declined apace with increasing difficulties in supplies, and with an ever-narrowing nutritional margin. The war cookery books increasingly dealt with special subjects, e.g. substitute foodstuffs like swede turnips and wild vegetables. The share of general recipe collections declined.

Cookery books are a specific kind of source material for historical research. Although they have been published in large numbers, especially since the second half of the nineteenth century, they are evidence only for the potential spectrum of dish and meal preparation and arrangement, since people could not be forced to follow the suggestions given by the authors.[5] The great number of war cookery books and nutritional guidelines, particularly in the first years of the war, proves that there was need for instructions for dealing with the change in nutritional conditions, and with less common foodstuffs, or even with foodstuffs which had already fallen into oblivion, e.g. dried fish.

Early war cookery books from 1914–15, as well as renowned nutritionists, pointed at a seemingly easy way to compensate for the shortage of meat and bread supplies: milk and dairy products combined with shredded cereals with which they were traditionally connected could provide the necessary protein.[6] However, this suggestion was based on false estimations of future production and distribution problems.

The argument used by the propagandists to increase the consumption of milk and cereal dishes, and of hitherto allegedly less popular dairy products such as skimmed milk, buttermilk, whey, and also sour milk products such as curd and curd cheese, was the exhortation to 'the resort to the 'plain customs of the forefathers', supported by national sentiment. Supposed and real foodways' traditions played an important part in the food propaganda. The resort to the plain customs of the forefathers' included the recommendation of morning and evening soups either with milk, water, or broth as basic liquids, and of milk and cereal mashes. These foodstuffs were supposed to compensate for the shortage of bread, meat and sliced cold meat in a way that was both economically and nutritionally appropriate. In 57.2 per cent of the forty-two war cookery books with general recipe compilations I have examined (see App. I), the foodways' traditions are used as a means of propaganda—a phenomenon, by the way, which also occurred in connection with the food shortages of the Second World War.

In the war cookery book recipes, the resort to traditional dietary patterns becomes evident, not so much in the preparation of certain dishes, as in their attribution to certain meals. Recipes for all kinds of milk soups, water soups or beer soups with different forms of cereal garnish, or for milk and cereal mashes, are no novelty in cookery books, but what is characteristic of the war period is that these dishes are so often definitely classified as forming morning and evening meals instead of the former coffee and sandwiches. Furthermore, soups and mashes form a relatively high percentage of the content of the war cookery books. The analysis of several such cookery books established that 20 to 30 per cent of all recipes were milk and cereal dishes, such as soups or mashes and sour milk dishes,[7] and curd and buttermilk dishes which had also been integrated into the propaganda. In cookery books from northern and eastern Germany, there are occasional hints as to the advantages of the southern German cuisine which had been traditionally characterised by dishes containing flour and cereals, and by having soups and mashes for breakfast.

Regional differences in the preparation of milk and cereal dishes are evident in the war cookery book recipes: authors from northern and eastern Germany tend to emphasise the preparation of milk and cereal dishes such as soup or mash, although books which claim to be non-regional try to offer a wider selection. Cookery books from southern Germany and Austria, on the other hand, integrate the numerous soups and mashes into their recipe collections, but they also boast a much broader and basically different choice of preparations, particularly for mushy dishes. In order to ease the step backwards towards so much mash, the dishes could be prepared in a variety of forms: i.e. *au gratin* in the oven or in the frying pan as rissoles or pancakes (*Schmarren*); the addition of eggs or gelatine turned them into solid, pudding-like dishes, and they could also be made into soufflés and so on. For milk and cereal dishes as well as for vegetable dishes—particularly for potatoes and swede turnips—the variety of preparations increased noticeably, but the traditional milk and cereal mash, or water and cereal mash, still formed the basis of the suggested diet. The diversity in the preparation of cereal dishes reflects the traditional cuisine of southern Germany, although the dishes consist of hitherto less familiar milled products such as corn, buckwheat, millet, pearl barley, groats and oatflakes.

In view of the milk and cereal-based traditional cuisine of this area, propaganda for the re-integration of morning and evening soups into the meal system is less often expressed explicitly in the prefaces of war cookery books from southern Germany. Only five (or 26.3 per cent) of the nineteen war cookery books of southern German and Austrian provenance include recommendations to increase the consumption of milk and milk products, or to re-integrate soups and mashes in breakfast and dinner menus; on the other hand nineteen (or 82.6 per cent) of the twenty-three cookery books from the northern, eastern and the central parts of Germany include such recommendations. One reason for these regional variations might be the fact that morning soups were quite common in southern Germany—at least in rural areas—until the 1930s.[8] Furthermore, southern Germany seems to have been better supplied with milk and dairy products—at least during the first years of the war.

Even the authors of cookery books could not help noticing the discrepancy between the stated increased consumption of milk, and the decline in milk production which occurred at the same time. They offer two solutions to this problem: the preferential use of skimmed milk, buttermilk or diluted full-cream milk, and the use of curd and curd cheese. Unfortunately, the retail trade did not offer a sufficient quantity of the much advertised skimmed milk, particularly at the beginning of the war. Skimmed milk was primarily used as animal feed, and many cookery books and nutritional guidelines counter this drawback with the terse advice to ask for skimmed milk as frequently as possible; this would necessarily result in an increased supply.[9]

Alongside the regional differences, there is a marked temporal development in the propagation of milk and cereal dishes. There is not only a decline in the number of war cookery books since 1916, but also an increasing integration of the swede turnip, a typical constituent of later war-time nutrition, and of a multitude of industrial products and substitutes into the diet. In their recipes, many authors allow for the increasingly problematic milk supply, particularly in the cities; towards the end of the war, only infants, small children and pregnant and nursing mothers could be provided with sufficient quantities.[10] More and more manufactured substitute products made their appearance in the recipes, and furthermore, it was now only very seldom recommended to increase the consumption of milk, particularly of skimmed milk, buttermilk and curd. Guidebooks which really wanted to offer pragmatic support, no longer commented on the increasing food shortage with morale-boosting slogans. They still recommended soups and mashes, but with increasing problems with supplies, these turned—in the guide books as well as in reality—into watery pearl barley or oat flake soups. Because of the lack of full-cream milk, fat and sugar, the only means of improving their taste was to add some diluted skimmed milk, evaporated or dried milk, fruit or saccharin.

THE NUTRITIONAL REALITY: CONSUMPTION DATA

It is quite likely that even during the first years of the war the daily cuisine could not include the variety of preparations proposed only in the more realistic nutritional guidelines and cookery books. Contemporary statements about the monotony of potato and turnip dishes, and complaints about the eternal humdrum of soups, are proof of this. Further evidence is the fact that the suggested variety of preparations required not only the basic constituents, in this case milk and cereals, but also many other scarce ingredients: fat for the rissole or pancake, eggs for the pudding or the soufflé, and sugar for the refinement of the taste. But these were goods in short supply, which probably made it fairly difficult for the majority of the population to follow such cooking instructions, particularly during the last two years of the war.

The source material for a nutritional evaluation of the First World War period is scanty both in quantity and quality; there is neither sufficient data on the development of the consumption of certain foodstuffs, nor enough menu-lists or autobiographical evidence pertaining to private households available. In addition, the statistics of the War Committee for Consumer's Interests (KAKI), with the exception of 1916 (April and June), cover only one month (April) for each of the years 1917 to 1918. Consequently, seasonal variations in food supply and consumption are not realisable at this stage. In addition, although the consumption data was decoded according to income and occupational groups,[11] regional and urban-rural differentiations—vital factors in considering the 'differential' consumption in times of crisis—which cannot be recreated in retrospect, are also lacking.

However, the KAKI statistics show a drastic decline in milk consumption as the war progressed; it was only 59.7 per cent of the imperial average of the years 1912–1913 in 1916, and declined to only 29.8 per cent of normal consumption in peacetime in 1918. The oft-publicised increase in milk consumption, especially for the years 1914–1915, through the use of skimmed milk and butter milk, is out of the question, at least for the second half of the war, as far as the imperial average is concerned.[12]

Regional differences in milk consumption were not only significant in times of peace,[13] but also, or perhaps especially so, in times of war. This latter situation can be verified by means of some surviving war statistics. The statistics of the KAKI show that thirty Bavarian families included in the general survey (the only regional survey which has survived),[14] compensated in part for the lack of nutritional products like bread and meat by using milk and milk products. The consumption of milk and cheese in these Bavarian households was on average above the standard of Pre-war consumption (see Table 5.1).

Proof of the special position of southern Germany as a region with traditionally high milk production *and* consumption rates, is known from circumstances other than the KAKI statistics, i.e. the milk consumption of the City of Munich. Although having declined from an annual 154.6 litres per capita in 1913 to 95.2 litres in 1918,[15] still the milk consumption of a Munich citizen in 1918 was twice as high as the imperial average.

The analysis of six Bavarian single household budgets from the years 1914 to 1916 presented in Table 5.2 also illustrates that even in the urban areas in the countryside, the milk supply of working-class families was considerably better, and that the southern German cuisine was rich in flour and milk, at least during the first years of the War.[16] Indeed, it is clearly evident from Table 5.2 that these Bavarian families maintained a sufficient standard of living from the nutritional point of view up to 1916, although food habits partly had to be changed.[17]

AN INDIVIDUAL VIEW OF REALITY: AUTOBIOGRAPHICAL TESTIMONIES

Consumption data alone do not give any information about actual consumer behaviour in relation to the suggestions given in the cookery books about kinds of preparation and meal arrangements, or especially about the processes whereby the traditional milk and grain soups or

Table 5.1 Bavaria and the Imperial average in April 1917 and 1912/13 (monthly consumption (kg) per consumption unit, 1912/13 per adult)

	Empire 1917[1]	Bavaria[2]	Empire 1912/13[3]
bread/rolls	8.407	7.189	11.010 (grain/baker's
flour/groats etc.	2.205	2.159	ware total)
potatoes	10.934	6.240	16.380
butter/fat	0.539	0.490	0.750
meat	1.983	2.061	4.020
fish/smoked fish	0.404	0.265	0.690
eggs	7 pieces	6 pieces	0.510 g
milk (l)	8	15	13.813 g (including milk-tins)
tinned milk			0.052
cheese/curd	0.334	0.819	0.360
vegetables/fruit incl.			
tinned vegetables	2.853	missing	10.740
sugar	0.897	1.092	1.410

Sources:[1] A. Roerkohl, *Hungerblockade und Heimatfront. Die kommunale Lebensmittelversorgung in Westfalen während des Ersten Weltkrieges*, Stuttgart 1991, 358;[2] F. Patschoky, 'Untersuchungen über die Lebenshaltung bayerischer Familien während des Kriegs. I: Die Erhebung des Kriegsausschusses für Konsumenteninteressen über die Lebenshaltung im Kriege', *Zeitschrift des K. Bayerischen Statistischen Landesamts*, **50**, 1918, 42–58, here 50;[3] C. von Tyzska, 'Hunger und Ernährung. Wirtschaftlicher Teil', in A. Gottstein and A. Schloßmann and L. Telekky, *Handbuch der Sozialen Hygiene und Gesundheitsfürsorge*, **5**, Berlin, 1927, 318–373, here 347.

54 Milk and milk products

Table 5.2 Monthly per capita consumption of milk, milk products, bread and grain in Bavarian households 1916 in comparison to the Imperial average (Engel unit, for the German Empire per consumption unit, kg)

foodstuff	fam. 1	fam. 2	fam. 3	fam. 4	fam. 5	fam. 6	Empire
milk (1)	24.7	21	16.5	16.0	13.7	14.9	8
cheese		0.242	0.249	0.211	0.485	0.246	0.363
butter	0.645	1.349	0.486	0.381	0.468	0.493	
butter/fat	0.888	1.378	0.529	0.729	0.657	0.653	0.862
bread/rolls (pieces)		13.4	12.5				
bread/rolls (lb)	11.022	6.621	10.088	6.228	13.319	14.670	8.304
flour	1.993	1.735	0.773	2.535	0.886	1.099	
rice	0.127	0.004	0.057	0.067	0.053	0.038	
semolina	0.283	0.128	0.219	0.157	0.120	0.111	
barley	0.105	0.015	0.354	0.067	0.070	0.019	
corn/millet	0.310	0.008	0.011				
oatmeal				0.067	0.092		
sago					0.103		
farinaceous food		0.223	2.047	0.023	0.224	0.043	
grain products (without bread and rolls, but including sago)							
altogether	2.818	2.113	3.461	2.916	1.548	1.310	1.445
meat and potatoes							
meat	1.984	1.068	2.686	2.725	1.462	0.625	1.901
potatoes	12.795	3.660	17.250	3.253	19.617	4.908	16.793

Sources: F. Patschoky, 'Untersuchungen über die Lebenshaltung bayerischer Familien während des Kriegs. II: Ganz-beziehungsweise mehrjährig geführte Wirtschaftsbücher', *Zeitschrift des K. Bayerischen Statistischen Landesamts*, **50** 1918, 592–626; A. Roerkohl 1991, 358.

mashes, might be substituted for the coffee and bread breakfast—a key aspect of the nutritional propaganda.

Apart from cookery books and household budgets, a third kind of source may allow us a deeper insight into individual food consumption, and the individual perception of, and solution to, changing nutritional conditions. Autobiographical testimonies in some cases include more or less detailed descriptions of the living standard and food habits of their authors. Although very individualistic testimonies, showing the conditions of life through the mirror of subjective perception and remembrance,[18] autobiographical accounts can give us an impression of a certain reality, especially those parts of them which only describe simple facts like localities, names or data.

Only a few examples of autobiographical testimonies from the war period give sufficient information to enable us to reconstruct the processes whereby cereal-based dishes replaced the coffee and bread breakfast, in addition to their numerous descriptions about hunger. The war diary of the Nuremberg trader's son, Ernst Buchner, begun in 1914 and ending with the termination of war in 1918, is such an example.[19] Buchner did not only record the hunger of his family living in southern Germany—which tended to be better supplied—at the beginning of 1916, but also registered the replacement of the coffee and bread breakfast by means of the traditional morning soups prepared as brown roux soups, or—with modern elements, but traditional preparation—as oatflake soups, at least for the children.[20] Buttermilk still came on his family's table two to three times a week in 1917, again prepared as a soup.[21] However, cheese, just like eggs, was a scarce commodity, which was not available in Nuremberg shops. A comparable situation can be proved for a family in Hamburg, which had the advantage of being able to exploit a piece of their own land. At the end of 1916 and the beginning of 1917, the mother only irregularly obtained a quarter or half a litre of skimmed milk for her two children, for which she had to queue for hours: 'It isn't worth queueing up, but the children are happy when they can get a little soup.'[22]

Proof for the repenetration of grain soups—at least in the area of child nutrition—can also be found in other autobiographical testimonies. A coal miner from the Ruhr District remembers: 'One wasn't even allowed to grind grain oneself. But we all know how things went . . . We children had to rough-grind secretly, rye to rye groats. With this and a little milk and saccharin—there was no sugar available—mum cooked a little soup. Later we were all fed up with these soups.'[23]

In spite of these examples, contemporary investigations prove that in September 1917 the urban poor of Berlin, and also the inmates of two Berlin hospitals, had a preponderance of coffee substitute and bread for breakfast, and only occasionally grain mashes and soups for dinner; in the hospitals grain soups were also had for lunch.[24] On the other hand, morning and evening grain and milk soups and mashes, in combination with large quantities of potatoes, sour milk products and curd for lunch and supper, were typical of the food habits of agricultural workers in the area around Berlin, even during the war. For these agricultural workers the traditional dominance of grain and milk dishes, apart from potatoes, in their diet in times of peace, continued during the war because of the better food supply in rural areas.

CONCLUSION

Available wartime consumption data and autobiographical testimonies indicate that a key aspect of the nutritional propaganda, occurring especially in northern and eastern German cookery books, namely, that nutritional deficiencies could be compensated for by reverting to milk and cereal-based dishes in morning and evening meals, was not, and largely could not, be fully realised, because of the milk-supply shortages. Occasional autobiographical references also show that where the reversion was made, it was usually for the benefit of children. The reality of the situation was that it was potatoes and vegetables, rather than curds and other milk products, which compensated for the lack of flour, meat and fat. It was only in some regions of southern Germany that urban families could balance the monotony of a potato diet by managing to keep the traditional southern German flour and milk-based products in their diet, at least until 1916–1917. Rural regions in northern Germany also had a better food supply and managed to retain the traditional dominance of milk and cereal dishes in their diet.

It is also evident that the tendency was to retain, if at all possible, the typical coffee and bread breakfast, even in a much reduced and altered form.

The use of industrially produced substitute foodstuffs was also a feature of the wartime cuisine. Even where there was a partial return to traditional milk and cereal dishes, these now included artificial ingredients such as egg substitute and dried milk.

The wartime consumption data is also of interest in that it brings into focus the regional variations of, and responses to, the food shortages, and also the difference in food supply between urban and rural areas. They also highlight the existence side by side of basically different food habits, such as the coffee and bread-based diet of the urban poor and in the hospitals of Berlin, as against the dominant milk, cereal and potato diet of the agricultural workers in the areas around Berlin.

In relation to wartime cookery books, the analysis shows that, during the first years of the war, there was a broad discrepancy between the normative suggestion in the cookery books and the realities of the food supply. During the latter part of the war, however, they were more in tune with the nutritional situation, especially in relation to the use of substitute foodstuffs and their preparation.

These cookery books from the First World War are also an important source of information for regional foodways in that the authors closely described the food habits of their own areas—cookery books from southern Germany and Austria, for example, include a broader range of mushy cereal-type dishes, while those of northern and eastern Germany, focus instead on simple milk and grain soups and porridge. In fact it is probably true to assert that cookery books from the First World War period—and perhaps from times of crisis in general—reflect more closely the realities of the nutritional situation, and also regional food habits, than the large number of cookery books published in times of peace.

NOTES AND REFERENCES

1 A. Roerkohl, *Hungerblockade und Heimatfront. Die kommunale Lebensmittelversorgung in Westfalen während des Ersten Weltkrieges*, Stuttgart, 1991. For further data see V. Ullrich, *Kriegsalltag. Hamburg im ersten Weltkrieg*, Hamburg, 1982; for specific aspects see G. Mai, *Kriegswirtschaft und Arbeiterbewegung in Württemberg 1914–1918*, Stuttgart, 1983; A. Lüdtke, 'Hunger, Essens- 'Genuß' und Politik bei Fabrikarbeitern und Arbeiterfrauen. Beispiele aus dem rheinisch-westfälischen

Industriegebiet, 1910–1940', *Sozialwissenschaftliche Informationen für Unterricht und Studium* **14**, 2 (1985), 118–126.
2 Roerkohl, *op.cit*, 200–216.
3 For a preliminary survey of war cookery books see R. Wulf, *Der heruntergekommene Lucull. Was Kochbücher und Ratgeber in Kriegs- und Nachkriegszeiten empfahlen*, Dortmund, 1990; Roerkohl, *op. cit*.
4 Of the 181 cookery books bearing the year of publication which I have found so far, ninety were published in 1915, while only twenty-one were published in 1918.
5 For a critique of printed cookery books as sources see H. Dose, 'Die Geschichte des Kochbuchs—Das Kochbuch als geschichtliche Quelle', in *Beruf der Jungfrau. Henriette Davidis und Bürgerliches Frauenverständnis im 19. Jahrhundert*, Oberhausen, 1990, 51–70; E. Hörandner, 'The Recipe Book as a Cultural and Soico-Historical Document. On the Value of Manuscript Recipe Books as Sources' in A. Fenton, T.M. Owen (eds.), *Food in Perspective*, Edinburgh, 1981, 119–144.
6 Suggestions like these were also part of official guidelines in an effort to change the food habits in accordance with realistic conditions and problems in food supply; for Westphalia see the examples in Roerkohl, *op. cit.*, especially 368, 369.
7 The analysis is based on the 42 war cookery books listed in App 1.
8 Cf. G. Wiegelmann, *Alltags- und Festspeisen. Wandel und gegenwärtige Stellung*, Marburg, 1967.
9 See for example [Carl] Oppenheimer, 'Magermilch in die Küche' in *Kleine Beiträge zur Volksernährung. Gesammelte Mitteilungen aus der 'Kriegskost'* I Berlin, *c.* 1914–15, 21–22.
10 For examples see Roerkohl, *op. cit.*.
11 On the differential consumption during the First World War see A. Triebel, 'Soziale Unterschiede beim Konsum im Ersten Weltkrieg und danach—Bruch mit der Vergangenheit' in T. Pierenkemper (ed.), *Haushalt und Verbrauch in historischer Perspektive. Zum Wandel des privaten Verbrauchs in Deutschland im 19. und 20. Jahrhundert*, St. Katharinen, 1987, 90–122.
12 Detailed data in Roerkohl, *op. cit.*, 358.
13 Even before the First World War, the milk consumption in the cities differed substantially, see W. Beukemann, 'Der Milchverbrauch der Städte. Sein Zusammenhang mit den Landesverhältnissen und dem Wohlstande der Bewohner' in *Bericht über die Allgemeine Ausstellung für hygienische Milchversorgung im Mai 1903 zu Hamburg*, Hamburg, 1904, 95–130.
14 F. Patschoky, 'Untersuchungen über die Lebenshaltung bayerischer Familien während des Kriegs. I: Die Erhebung des Kriegsausschusses für Konsumenteninteressen über die Lebenshaltung im Kriege', *Zeitschrift des K. Bayerischen Statistischen Landesamts*, **50**, 1918, 42–58.
15 E. von Reitzenstein, *Die Milchversorgung der Stadt München. Ein Beitrag zur oberbayerischen Milchwirtschaft*, Kempten 1930, 46.
16 For statistical details of the annual consumption see F. Patschoky, 'Untersuchungen über die Lebenshaltung bayerischer Familien während des Kriegs. II: Ganz- beziehungsweise mehrjährig geführte Wirtschaftsbücher', *Zeitschrift des K. Bayerischen Statistischen Landesamts*, **50**, 1918, 592–626.
17 This estimation in Patschoky (1918, II) who did a differentiated analysis on these household budgets. Unfortunately as there are no comparable detailed household budgets surviving from other regions of the German Empire or from rural households for the war period, the analysis had to be confined to the comparison of the Bavarian situation with the Imperial average.
18 For the perception of situations of hunger especially in the autobiographies of workers see Lüdtke, *op. cit*.

58 *Milk and milk products*

19 E. Buchner, *1914–1918. Wie es damals daheim war. Das Kriegstagebuch eines Knaben*, Nürnberg, 1930.
20 *Ibid.*, 111.
21 *Ibid.*, 171, 174.
22 Ullrich, *op. cit.*, 95.
23 *Hochlarmarker Lesebuch. Kohle war nicht alles. 100 Jahre Ruhrgebietsgeschicht.* Oberhausen, 1981, p. 62.
24 A. Loewy and C. Brahm, 'Untersuchungen über Art und Wirkung der Kriegsernährung', *Zeitschrift für physikalische und diätetische Therapie*, 23 1919, 169–194.

APPENDIX 1
List of Wartime Cookery Books

Anleitung zum Wirtschaften in der Kriegszeit. Zugleich ein Kriegs-Kochbuch, Wiesbaden, 1915.
Barth, Marie/Laura Barth, *Die Milch und ihre Verwendung im Haushalt. Ein Beitrag zur Lösung der Ernährungsfrage mit einer Sammlung erprobter Rezepte*, 4. ed., Aarau, 1916.
H. Behnke, *Zeitgemäße Küche in dritten Kriegsjahre 1916–17*, 2. ed., Hamburg, 1917.
Berta Dißmann, *Kriegsmahlzeiten. Vorschläge für nahrhafte und zeitgemäße Tagesbeköstigung mit einem Anhange über Einmachen, ferner über Herstellung, Wert und Gebrauch von Kochkiste, Kochbeutel und Kochrucksack, Juni des Kriegsjahrs 1915*, Dresden, 1915.
Berta Dißmann, *Wer spart, hilft siegen! Grundlagen zur Herstellung einer billigen und nahrhaften Kost*, Dresden, 1915.
Erfahrungen der Kriegszeit. Eine Ergänzung zum großben Kochbuch des Schwäbischen Frauenvereins, Stuttgart, 1918.
Henriette Fürth, *Kriegsküche für jedermann*, Frankfurt am Main, 1916.
Marianne Götze, *Sächsisches Kriegskochbuch. Anweisungen für eine sparsame und kräftige Ernährung*, Leipzig, 1915.
Mary Hahn, *Kriegs-Kochbuch*, Wernigerode, 1916.
Luise Hainlen, *Schwäbisches Kriegskochbuch*, 3. ed., Stuttgart, 1916; 7. ed., Stuttgart, 1918.
Caroline Herrmannsdorfer, *Hausfrauensorgen in Zeiten des Krieges und der Teuerung. Ratschläge zur Herstellung billiger und nahrhafter Hauptmahlzeiten, einschließlich unbekannter Originalkochrezepte für den Haushalt nebst Anweisung zur einzig richtigen Zubereitungsart des Polentamehls als Fleischersatz*, Munich, 1915.
Hedwig Heyl, *Kleines Kriegskochbuch*, Berlin, 1914.
Hedwig Heyl, *Kriegskochbuch. Anweisungen zur einfachen und billigen Ernährung, Ausgabe für Süddeutschland*, 54. ed., Berlin, 1915.
Hedwig Heyl, *Kriegskochbuch. Anweisungen zur einfachen und billigen Ernährung, Ausgabe für Norddeutschland*, 67. ed., Berlin, 1915.
Hedwig Heyl, *Kriegskochbuch. Anweisungen zur einfachen und billigen Ernährung. Ausgabe für Ostdeutschland*, 94. ed., Berlin, 1915.
Louise Holle, *Davidis-Holle. Praktisches Kriegskochbuch*, 3. ed., Bielefeld/Leipzig, 1916.
Kochbuch für den einfachen Haushalt, Kriegsausgabe, 4. ed., Breslau, 1915.
Kochbuch für die Kriegszeit, Dortmund, 1915.
Kriegskochbuch, Kattowitz, 1915.
Bayerisches Kriegskochbüchlein. Anweisungen zur einfachen und billigen Ernährung unter besonderer Berücksichtigung ländlicher Verhältnisse, 2. ed., Munich, around 1914–15.
Deutsches Kriegskochbuch und Anleitung zum Gemüsebau für den eigenen Haushalt, hg. v. irgend einer Deutschen, Jena, 1915.

Kriegskochbuch der Frauenhülfe. 50 zeitgemäße Rezepte, Potsdam, 1915.
Kleines Hamburger Kriegskochbuch, Hamburg, around 1915.
Neue Kochvorschriften. Gesammelt im dritten Kriegsjahre vom Allgemeinen Deutschen Frauenverein in Graz, Graz, 1917.
Kriegsküche 1917. Von den Haushaltungslehrerinnen des Schulkreises Schopfheim, Waldshut, 1917.
Anna Larisch, *160 Kochrezepte für die Kriegs-Zeit*, 2. ed., Brünn, 1915.
V[ictoria] Löbenberg, *Das deutsche Sparkochbuch für Kriegs-und Friedenszeit mit Gesundheits und häuslichem Ratgeber*, Munich, 1916.
Nachtrag zum Kriegskochbuch. Vier Wochenspeisezettel, Stuttgart, 1915.
Josephine Nagel, *Die Neue Kriegsküche. Kochanweisungen für fleisch-und fettsparende Tage*, Berlin, s.d.
Josephine Nagel, *Die Kriegsküche im Jahre 1916. Kochanweisungen für fleisch- und fettsparende Tage*, Berlin, 1916.
Meli Nefflen, *Kochbüchlein für die Kriegszeit. Eine Sammlung bewährter und erprobter Rezepte*, Ebingen, 1915.
Mathilde Nouvel, *Fleisch- und fettarme Kriegskost. Gerichte aus Kartoffeln, Gemüse und Obst nebst Kinder- und Krankenkost für Stadt und Land*, Breslau, 1916.
El[isabeth] Peschges, *Norddeutsche Küche in der Kriegszeit, Mönchen-Gladbach, 1915.*
Gertrud Pick, *Volkskriegskochbuch. Anweisung zur kräftigen und billigen Ernährung in 100 ausprobierten Rezepten*, Gotha, 1915.
Marie Priester, *Spart Fleisch und Brot zur Zeit der Not. Kriegskochbuch für fleischfreie Tage. 200 Mittag- und Abendessen mit 92 Kochvorschriften*, Frankfurt am Main, u.d.
Rezepte der Kriegskochkurse des Vereins wirtschaftlicher Frauentätigkeit in Stadt und Land e.V. Marienheim Speyer, Speyer, 1916.
Betty Roß, *Kriegsküche. 150 ausgeprobte billige und schmackhafte Speisen. Ein Sparbuch für jede Hausfrau*, Lorch, u.d.
M[aria] Schneider, *Rheinische Küche in der Kriegszeit*, Mönchen-Gladbach, 1915.
Toni Schroeder, *Notstandsküche. Kochanweisungen*, Berlin, 1915.
Emma Wiedermann, *Das Kriegskochbuch des Vaterländischen Frauenvereins*, Berlin, 1915.
Emma Wundt, *Badisches Kriegskochbüchlein. Winke für die Hausfrauen während der Kriegzeit*, Karlsruhe, 1915.

Part II
The Urban Retail Milk Trade

6 The supply and sale of milk in nineteenth-century Edinburgh

Una A. Robertson

Mention the name of actor Sean Connery in Edinburgh and someone invariably recalls that in his early days he was a milkman with the St Cuthbert's Co-operative Society, delivering bottles of milk from a horse-drawn float.

This snippet of local history made me wonder at what point milk had been delivered direct 'to the door', as against the generally-held picture of milk sold by peripatetic milkmaids, or direct from the cow. Instead of the expected clearcut divide between an older, discredited system and the modern hygienic one enforced by legislation, the story proved far more complex than anticipated and involved what might be termed 'a cast of thousands'—including Scotland's very first Medical Officer of Health, a Sunday School teacher and the Russian Czar, the editor of *The Scotsman* newspaper, plus assorted Scottish landowners, and the many visitors to a certain pleasure garden, now beneath a Glasgow thoroughfare, where fresh milk and strawberries were sold. That milk is central to our story.

Early in the nineteenth century milk and milk-based products were commonplace in Edinburgh. Domestic account books include many references to them and locally-published cookery books contain a plethora of recipes using milk or cream. Sometimes included is a local delicacy called 'Corstorphine Cream'—named for a village then outwith the City, though nowadays a suburb—and the evocatively-named 'Hatted Kit'.[1] Agricultural manuals discussed the rival merits of town dairies, and some mentioned Edinburgh, while a high consumption of milk, cheese and buttermilk by the population generally was revealed in the 1790s Survey of Scotland. In Edinburgh, butter, cheese, curds and whey were sold along the High Street, now better known as the Royal Mile, and it was estimated that during the four summer months over £1000 was spent on butter-milk (or *sour dook*), at the rate of one old penny per Scots pint, brought into town by milkmaids on horseback, the barrels strapped up behind them.[2]

By this time it was mostly cow's milk that was used here and the full-milk Dunlop cheese developed in Ayrshire had superseded the earlier skimmed and goat's milk cheeses. Those citizens wishing to drink goat's milk or whey for reasons of health had to travel to spas such as Dunkeld or Moffat where it was available during the summer, and even though Edinburgh had its own healing waters goats do not seem to have featured in 'the cure'. However, ass's milk, considered beneficial for bronchial troubles, was sold until the 1830s, but if it was unavailable an 'Artificial Ass's Milk' could be made according to recipes given in some cookery books.[3]

The fact that a considerable portion of the Royal Mile was given over to the sale of milk-based products would suggest a plentiful provision. What is not plentiful, though, are details to show how Edinburgh was actually supplied with these commodities, the word 'supply' here covering both the production of milk from the cow, as well as its subsequent availability. The selling of sweet (i.e. fresh) milk is rarely mentioned. The raucous 'cries' of many street traders are described, but milk is not one of them,[4] and portraits were drawn of certain traders going about their business but, again, the seller of milk was not included.[5] One wonders why. Although it is to be hoped that the way milk was sold in Edinburgh was far removed from the nauseous description given by Smollett for London,[6] the suspicion remains that, maybe, the sight was not for the fastidious! Several nineteenth century writers deplored the dearth of information generally, and although the situation in cities such as London, Liverpool and Glasgow is on record, little has been written about Edinburgh.

While the capital remained tightly-packed, straddling the ridge that runs between the Castle and Holyrood, milk and other foodstuffs were brought in from the surrounding area. A Town Council Minute of 1689 required '. . . that the inbringers of the milk may be in reddiness to remove out at the ports [i.e. gates] at 8 of the clock in the morning'.[7] Over the next hundred years or so, the supply situation remained basically unchanged even though Edinburgh had expanded greatly in the interim, and more especially after 1767 when the New Town began to spread over the fields to the north. However, between 1800 and 1850 the city's one square mile became seven, and its population rose from 90,000 to almost a quarter of a million; its area and population doubled again by the end of the century, resulting in some of the most spectacular slums in Europe, as well as the breakdown of traditional methods of supply.

Prior to this, fresh milk for a town was only considered profitable if produced within a radius of about two miles and, even for London, it was rarely transported above ten; butter and cheese came from further afield. As Edinburgh expanded over the surrounding farmland, so the number of dairy cows kept within the boundaries increased; in 1752 there were only two cowfeeders,[8] but by 1865 the figure had increased dramatically to 171 byres, housing some 2085 cows.[9]

Considerable confusion exists over the terms 'cowfeeder' and 'dairy', and although the terms are sometimes synonymous, historical usage also indicates that a 'dairy' tended to produce milk for others to retail, whereas a 'cowfeeder' worked on a smaller scale and sold the milk himself. Such an establishment was often termed 'a dry dairy' and some operated in Edinburgh until the 1930s, and for even longer elsewhere.

The preferred dairy cow came to be the Ayrshire which had various advantages over the former favourite, the Roxburgh short horn. The Ayrshire yielded an average of 11 English quarts of milk a day on moderately good fodder, and it fattened well for the butcher thereafter, though on those terms 'the Highland Scot', which fattened on next-to-nothing, might have been an even better buy! Cows for dairy use were generally bought in just before, or just after, calving and commonly the calf itself got short measure. It was considered best for calves to be born in the spring but matters could be arranged so as to ensure uninterrupted supplies of milk for the dairy. Milking took place twice a day and although mechanisation had been introduced for many dairy chores, devices for milking were rare and one person would take an hour to milk eight to ten cows. It was a common maxim that only a well-fed cow would produce good milk. While some dairies took pride in their cows, driving them out regularly to nearby pasturage for feeding and exercise, others kept their beasts tethered in stalls the entire time and fed them on absolute rubbish, a cause of grave concern. If poor feeding led to poor milk, then the meat quality—to be butchered was the fate of these animals once they ceased to be profitable in the dairy—was distinctly dubious. London's dependence on carcases from such animals fed entirely on oil-cake was deplored and the day was anticipated 'when the cattle so fed will be sent alive in steam boats to the Continent . . . where the taste of the inhabitants . . . is less refined'.[10]

Insalubrious conditions in milk production were not inevitable. Alternatives were not only possible, but had been practised to enormous acclaim some years previously, though here the story momentarily moves to Glasgow, Edinburgh's arch-rival in the west.

Recall, if you will, that pleasure garden mentioned earlier. This link in our story started in 1802 when an innovative weaver-cum-Sunday School teacher, William Harley, bought lands on the Sauchy Haugh, one mile north of Glasgow Cathedral, and developed the springs of water there, distributing it in pony carts throughout the town. Next to the reservoirs Harley set up public baths and laid out pleasure grounds, where folk might stroll afterwards, drinking fresh milk from a cow sent down from his own house, or taking cream with his finely-flavoured strawberries. The public tired of everything except the milk and thus 'Harleys byres' were born which remained 'a household word in Glasgow for several generations'.[11] The first byre, built in 1809, was for 24 cows but demand was such that it led to a vast complex housing up to 300 beasts, every one scrupulously fed, tended and groomed twice daily, as though they were cavalry horses. The buildings were kept equally spotless and numerous calves and pigs were reared on the surpluses. Byremen looked after the cows although, we are told, 'the milkers were generally women, their wages being half that of men, and they were found to milk as well'.[12] Initially, milk was sold at the dairy, but later it was distributed throughout Glasgow by well-appointed pony carts, harness and brass shining, the milk being sold from padlocked vessels and run through a stopcock.[13] The byres attracted so many sightseers that a viewing platform was constructed, with curtains which were flung open—to gasps of amazement all round. The future Czar Nicholas came, as did many other foreign princes, and when Harley bankrupted himself trying to do for bread what he had achieved for milk, the Czar invited him to establish such a dairy in Russia. Harley died *en route* for St. Petersburg in 1829.

The 'Willowbank System' (named after Harley's estate), or 'The Harleian System' (the title he gave his book), was considered original as regards the logical arrangement of the buildings, its effective system of management, and the emphasis on absolute cleanliness throughout. Its real originality, though, lay in the concept of pure fresh milk *in a town*! There were those who wished to establish the system throughout the country, seeing its essentials were relatively simple, but Harley's excellence went unchallenged. Outside London, for example, in still-rural Islington, two huge concerns were run by a Mr Rhodes and a Mr Laycock, while the smaller Metropolitan Dairy was in Edgeware Road. Slightly further afield, the East Acton Dairy was unusual, not only for its general superiority, but because 'it was projected, completed, and continues to be conducted by a lady'.[14] Management methods are detailed for each, but none attained Harley's standards, particularly as regards the quality of milk, which was always considered vulnerable where the milking was contracted out to retailers.

The 'Willowbank System' eventually came to Edinburgh and here we have Harley's own evidence. He advised the two parties who came to see him to combine for greater profitabilty, seeing they both intended to follow his methods, but two separate companies were set up. One took a farm and converted its buildings, while the Caledonian Joint-Stock Company expended £22,000 on buying the lands of Meadowbank and Wheatfield and built a magnificent cow-house 'which attracts and deserves the attention and admiration of visitors, and is well worthy the metropolis of Scotland'.[15] The new concerns caused an outcry for interfering with the old cowkeepers, but Harley believed they had forced the smaller dealers to raise their standards in order to keep their customers. As he proudly said: 'The City of Edinburgh is now better supplied with milk than at any former period'.[16] Brave words! The Caledonian Joint-Stock Company never achieved its potential and was soon let to a Mr Bellis who kept between 60 to 80 cows there instead of the 200 envisaged; but even this venture was short-lived and Edinburgh's model dairy vanished from the scene about 1832.

Some thirty years later, and two-thirds of the way through our chosen period, what do we find? The evidence suggests a deterioration due, mainly, to the ever-increasing size of the city. Dr Henry Littlejohn, appointed Medical Officer of Health in 1862, investigated various aspects of the city's health and part of his 'Report' concerned the 171 cowbyres he found, sited in every quarter. The majority he considered to be a severe hazard, both to those living nearby, and to the cows inside. Badly built, with inadequate ventilation and drainage, the byres were usually overcrowded and lacking in any provision for sick animals, which were invariably sold off to the slaughter houses for public consumption thereafter. While castigating the conditions in which the milk was produced, Dr Littlejohn believed, as had others before him, that 'poor milk', the focus of so many complaints, was caused by poor conditions rather than by anything more sinister and, equally, that no disease had been caused by milk in the previous ten years. He thought that in Edinburgh, as well as in London, little adulteration of milk occurred, apart from the addition of water, and that harmful ingredients were never introduced. This was in stark contrast to the malpractices of earlier years, detailed below.[17] His optimism was still valid in 1910 when 24 samples out of 439 were found to be impure, mainly due to the addition of water.

Although Dr Littlejohn found much to alarm him, the period between 1830 and 1865 had witnessed the beginnings of several radical developments that would profoundly effect the way in which large urban populations were provisioned and these were to continue through the later years of our survey and beyond. Legislation regulated standards of hygiene, railways were established and refrigeration techniques improved; in due course in due course, pasteurization, bacteriological research and other scientific developments were applied to the problems. The cumulative effect of these factors, along with others, changed the way foodstuffs were sold and contributed generally to the retailing revolution witnessed by the nineteenth century, whereby the retailer became less important while the organisation of the retail trade became vastly more complicated.

From this time onwards the history of Edinburgh's milk supply scarcely differs from that of any other large city. However, certain features were of particularly local relevance: in the first place, Dr Littlejohn single-mindedly fought the cause of public health for fifty-two years and his influence lasted for considerably longer. Legislation relating specifically to town dairies and milk came into force at national level, and was then adapted to local needs by the town council: Acts of particular significance included the *Cattle Sheds in Burghs (Scotland) Act, 1866*; the *Contagious Diseases (Animals) Acts, 1878 and 1886*, superseded by the *Diseases of Animals Act 1894*; the *Dairies, Cowsheds & Milk Shops Order, 1885* and its *Amending Order* of 1887, and *The Public Health (Scotland) Act, 1897*. This legislation was then incorporated into relevant Edinburgh City Acts.[18]

The thou-shalt-nots enumerated therein are not for the easily nauseated. These bye-laws were strictly enforced and inspectors gradually appointed. In 1879 regulations covered ice-cream dealers and in 1896 registration of ice-cream shops was sought. The findings on the seventy-eight shops registered that year make for disturbing reading.

Stricter controls over milk went hand-in-hand with preventing the meat from diseased cows being sold for public consumption.[19] Certain cattle diseases had been endemic for years and the new regulations, demanding instant slaughter and disposal, led to a shortage of cows in the town's dairies and, in the short term, to a greater proportion of milk coming from outwith the town and therefore not subject to city regulations! In the long-term, the number of 'dry dairies' showed a steady decrease. By this time, though, the railway networks ensured that large towns could be provisioned from almost any part of the country and, as early as December 1824, Charles MacLaren, editor of *The Scotsman* had correctly forecast their potential.[20] To take just one example; when the Scottish and English networks eventually linked together, the traffic in meat and fish sent the five hundred and fifteen miles from Aberdeen to London proved highly successful.

Dr Littlejohn was not the first to anticipate that the railways would obviate the need for town dairies altogether, yet he found that between 1855 and 1865 Edinburgh's cowkeepers had actually increased in number. He calculated that 132,000 gallons of milk, a mere seventeenth of the total, was brought in by the three principal railway companies.[21] The shortages of the 1860s stimulated rail transport for milk, which had been slow to gain acceptance due, presumably, to the fact that milk was still carried in open vessels and arrived looking more like dirty cheese! In time, the railway companies provided better facilities and every aspect of milk transport was investigated, which even included the holding of competitions to find better designs for containers and lids.

Meanwhile, the desire for purer foods and better quality shopping was being tackled from another angle. The railway workers living around Edinburgh's Haymarket Station provide yet another strand in our story, being instrumental in forming the St Cuthbert's Co-operative Society. Whether the movement originated with Robert Owen in New Lanark in the 1820s, or should be traced back to the Fenwick Weavers who banded together to buy oatmeal in 1769, is debateable. However arrived at, the new principles of 'vertical integration' and 'the branch system', developed by the Co-operatives, became the keys to large-scale trading and were subsequently practised to great effect by Glasgow grocer Thomas Lipton and others thereafter.

St Cuthbert's was inaugurated at a public meeting in 1859 and two months later their shop was trading in Fountainbridge. Initially, only groceries were stocked; other goods were then added, with mixed fortunes, and in the 1880s two short-lived attempts were made to supply milk to members through contracts with local dairymen.[22] Not until the 1920s was land bought to maintain a 50-strong dairy herd; a creamery followed, then a pasteurization plant (in which the Co-operative Societies were pioneers), which doubtless helps to explain the local figures: whereas 43 per cent of milk sold in 1925 was bottled, this had risen to 84 per cent in 1936. The great days of the Co-operative Societies are outwith our period, as is the way grocery shops expanded their stock to sell fresh provisions alongside their customary dried goods, but by the 1950s St Cuthbert's had 100 milk floats and 100,000 customers. As recently as 1968 their head stableman said he could not envisage a future without horses to deliver milk.[23] They vanished off our streets a few years back.

Recent memories are of these horse-drawn milk floats which delivered bottled milk, yet many still recall earlier ways of buying milk, from a 'dry dairy', or from a pony and spring cart carrying galvanized churns or a cask, where the milk was ladled out or run through a tap. Memories would need to be long indeed to recall the street traders of the Royal Mile.[24]

If Sean Connery, born only a stone's throw from that first St Cuthbert's shop in Fountainbridge, is in the forefront of our story, then behind him looms the figure of William Harley, citizen of Glasgow. Jostling him in the shadows stand a host of local characters—all sellers of milk, or butter, or cheese, or buttermilk—and we must not forget 'Kirsty, nicknamed Rousty Kirsty' a seller of curds and whey, who was the last of her race to trudge through the streets of Edinburgh.[25]

NOTES

1 M.Dods *The Cook and Housewive's Manual*, Edinburgh, 1842, 394–5.
2 J. H. Jamieson, 'Edinburgh Street Traders and their Cries', *Old Edinburgh Club*, Vol. II, Edinburgh, 1909, 203; Scots £1 = 1s 8d sterling (pre-decimal) or £0.09 today.
3 *Gray's Annual Directory* 1833–4 ... for Edinburgh and its vicinity, Edinburgh, 1833, 134: 'Smith. William, ass milk dairy, Grove street'.
   Recipes: H. Glasse, *The Art of Cookery made plain and easy*, London, 1755, 239; M. Dods, *op. cit.*, 1842, 402.
4 *Old Edinburgh Club*, Vol. II, Edinburgh, 1909, *passim*.
5 H. Paton (ed.), *A Series of Original Portraits and Caricature Etchings*, by ... John Kay, 2 vols., Edinburgh, 1837.
6 T. Smollett, *The Expedition of Humphry Clinker*, Edinburgh, 1788, Vol. I, 136
7 *Old Edinburgh Club*, Vol. II Edinburgh, 1909, 202: 'In 1689 an Act was passed prohibiting the bringing of milk into the city on Sundays after 7.30 am, in order "that the inbringers of the milk may be in reddiness to remove out at the ports at eight of the clock in the morning."'
8 J. Gilhooley, *Directory of Edinburgh 1752*, Edinburgh, 1988, 97.
9 Dr H. Littlejohn, *Report on the Sanitary Condition of the City of Edinburgh*, Edinburgh, MDCCCLXV, 50.
10 J. Loudon, *An Encyclopaedia of Agriculture*, London, 1831, 1029.
11 J. Galloway, *William Harley: A Citizen of Glasgow*, Glasgow, 1901, 41.

70 Milk and milk products

12 W. Harley, *The Harleian Dairy System*, London, 1829, 85.
13 Harley, *op. cit.*, 90: He said that preventing the distributors from adulterating the milk 'cost more study than all the other departments of the Establishment; although it was only necessary to ascertain how to admit air to allow the milk to run off by the stop-cock, and to prevent water being put into the milk.'
14 Harley, *op. cit.*, 161.
15 Harley, *op. cit.*, 151.
16 Harley, *op. cit.*, 152.
17 Rev. J. M. Wilson (ed.), *The Rural Cyclopaedia*, Edinburgh, 1849, Vol. III, 456–7: A common malpractice was to remove the cream and then dilute the milk with water; an 'old trick' was to thicken this impoverished liquid with wheat flour, although this soon settled to the bottom; a better result was obtained when sugar or sugar candy was dissolved in the milk and the wheat flour was boiled; an emulsion of almonds or hemp seed could be added to diluted milk while the addition of egg whites gave 'a rich and creamy consistency'; carbonate of magnesia, calcined or prepared chalk, and bad mutton fat were other substances sometimes introduced and starch was certainly used.
18 Sir T. Hunter, *Statutes and regulations affecting dairies and milk supply in Edinburgh*, Edinburgh, 1898, *passim*.
19 Edinburgh Magistrates Minute Book: 9/7/1847: a veterinary surgeon in Dalkeith reported that many cattle had died of some dread disease during the previous two years and that the carcases, while not sold publicly, were 'salted, made into rounds, hammered up into sausages, or made into pies'.
20 D. Bremner, *Industries of Scotland*, Edinburgh, 1869, 86–7.
21 Littlejohn, *op. cit.*, 72 Appendix VII: 'By the North British Railway (including *the Edinburgh, Perth & Dundee lines*) 11,206 gallons; By the Edinburgh and Glasgow Company 915 gallons; By the Caledonian Railway 120,000 gallons.'
22 W. Maxwell (ed.), *First Fifty Years of St. Cuthbert's Co-operative Association Ltd. 1859–1909*, Edinburgh, 1909, 140–1.
23 *Edinburgh Evening News*, 9/1/68: 'When the clip clop fades'.
24 *Old Edinburgh Club*, Vol. II Edinburgh, 1909, 182: 'The old cries appear to have continued unabated till about the sixties and seventies of last century, after which there seems to have been a gradual decline in their number.'
25 *Old Edinburgh Club*, Vol. II, Edinburgh, 1909, 203.

7 The retail milk trade in transition: a case-study of Munich, 1840–1913

Uwe Spiekermann

From the middle of the nineteenth century, industrialisation changed not only the sphere of production in Germany, but made further economic development dependent on increasing sales of consumer goods and an extended distribution system. Fundamental changes in German retailing had occurred which altered first of all the life-styles of urban populations. Advertising and industrially produced goods led to new consumer demands and the shop in its differentiated form became the centre of the fledgling consumer society.[1]

These general changes affected also the sale of goods which were not industrially produced. The foodstuff trade is a good sample of an indigenous development of retailing, and to exemplify the processes involved we shall consider the changes and development of the Munich milk trade during the years 1840 to 1913.[2]

1. The Munich retail milk trade did not begin in 1840, but rather at that time two developments occurred which were to remain dominant until the First World War: (a) the breaking of the direct relationship between consumer and producer, and (b) the establishment of the shop as the centre of the retail trade.

Around 1840 the milk supply in Munich was in the hands of three different groups: farmers, milkmen, and the so-called milk storehouses (*Milchniederlagen*).[3] The market trade in milk had already ceased in 1807, and the street trade was not very important. Farmers and milkmen produced according to demand and delivered the goods directly to their customers. For the milkmen, who came on the scene at the beginning of the eighteenth century, their milk business was their means of livelihood, while the farmers, who came from the surrounding areas, mainly sold their products to obtain cash for purchasing goods they could not produce themselves. The milkmen kept their cows within the Munich *Burgfrieden* and produced milk for sale only. They set up little street stalls to sell the excess milk, and also introduced the sale of milk from carts. This new form of competition induced some farmers, especially the larger ones, from the beginning of the nineteenth century, to store milk in rented

72 Milk and milk products

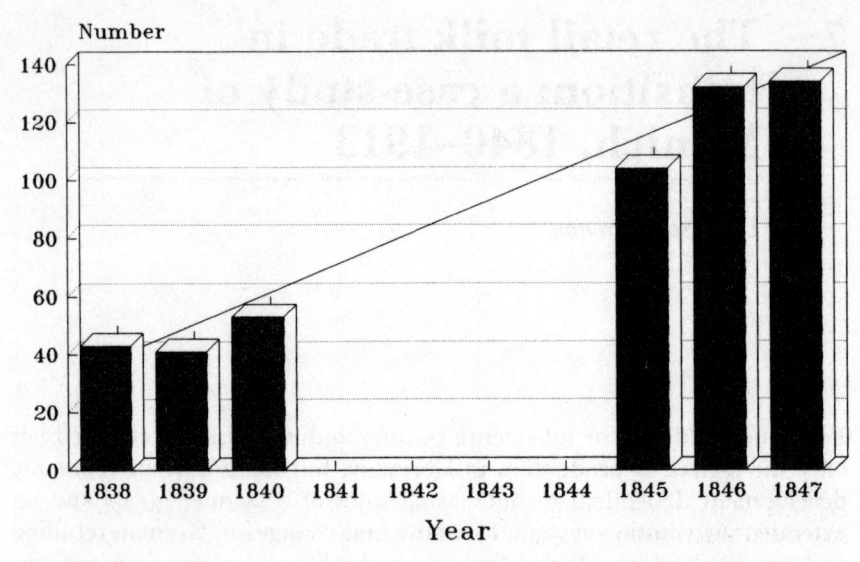

Fig. 7.1. Milk storehouses in Munich 1838–1847. (Source: StA München Gewerbeamt 5322)

business premises, and to have it transported from there by milkmaids (*Milchmädchen*) to their customers. These large farmers had simply to supply the milk to these so-called milk storehouses, and the retailing of it was taken over by their employees. Thus, the farmers were able to match the milk quality of the milkmen. Between producer and consumer the milkmaid appeared first as a saleswoman, then as a retailer, but from 1840 milk storehouses were increasingly independently rented by milk saleswomen supported by the producers who granted them a loan. Starting in the centre of the town, this first form of the shop gained more and more ground within the town.

2. *When milk became a purely commercial product the state authorities increasingly reacted with regulatory decrees and an intensification of control. For the retail trade the direct official control of milk played a less important role than the stringent regulations for the sale of milk. The pressures for rationalisation and professionalism in the milk trade occurred not through official control of milk, but rather through the shop and sales arrangements.*

Like all other agricultural items, milk was a free product in Munich being subject only to the general market and police regulations. Due to controls by local officials comparatively high food-stuffs standards were achieved long before the establishment of a chemical investigation office in 1880 or 1884.[4] There are no convincing sources for Munich indicating that adulteration of milk was common or usual.[5] Although complaints about unhygienic conditions in milk storehouses were often made by

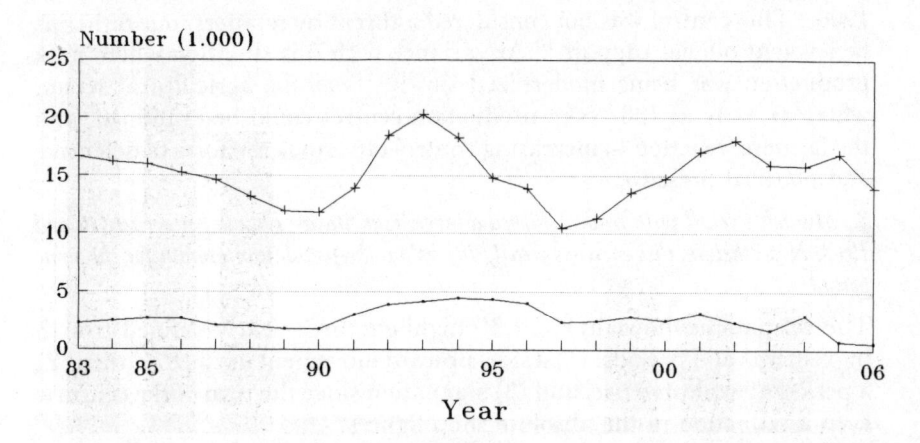

Fig. 7.2. Milk examination in Munich 1883–1906. (Sources: Marktverkehr, 1893, 369: Creuzbauer, 1903, 296; Jahresübersichten 1896, 67*, 1904, 24*, 1907, 30*)

milkmen, these complaints, serving the specific purpose of denouncing the intermediate storage stage of the milk by farmers, arose from business competition and were not substantiated in reality.

Munich Town Council (*Stadtmagistrat*) promoted direct relationship between producer and consumer until the early 1860s, and tried several times to re-establish smaller milk-markets to increase the ever diminishing number of farmers selling directly to consumers.[6] Munich was one of the first towns in Germany to regulate the sale of milk by means of a special decree in 1862.[7] To keep the number of milk storehouses in check, high hygienic standards for the shop, the adjoining rooms, and the sales equipment, were fixed. But despite these strict regulatory measures the significance of the shop trade was not diminished. The proprietors obviously reacted to this pressure for rationalisation and professionalism with an increase in sales and an improvement of the milk shop which was then experiencing its first flow of customers.

Any other reaction by the milk shop proprietors would have been fruitless in Munich, for not only were the legal norms tightened up later on, but also the actual control systems were improved decisively. Supported by physicians and chemical engineers from the university, the district supervisors exercised control using the most modern means available at that time.[8] But their main occupation was the inspection of the shops and sales equipment; every shop was controlled between ten and twenty-five times a year.

In Munich there was continual milk control, and through the tightening of the decrees in 1892, the national model role of Munich in this connection was underlined, as well as through later decrees about infant milk

74 Milk and milk products

(1899), or the reorganisation of the whole control of the food industry in 1906.[9] This control was not considered a threat by retailers, but rather as benevolent official support.[10] At variance with this situation where milk production was being modernized slowly[11] was the agricultural sector, where as early as 1887 even production centres could be examined, and the farmers' reaction to increasing control often took the form of defensive and political pressure.

3. *Munich's retail milk trade achieved a large quantitative growth between 1840 and 1913. Nevertheless, this increase clearly lay below the population growth for the same period.*

The estimations shown in Fig. 7.3[12] highlight three clearly-defined growth movements and periods: (1) a slow upward movement until 1875, then (2) a period of explosive rise, and (3) stagnation since the turn of the century, even a reduction in the absolute shop figures.

If this is brought into relationship with the quickly growing population (Fig. 7.4.), a clearly visible rise in the average number of customers of the retail milk trade since 1845, with the exception of the period between 1875 and 1885, becomes clear. Between 1840 and 1913 the average number of purchasers per shop had doubled (Fig. 7.5).

4. *During the period between 1840 and 1913 Munich probably experienced a duplication of the average milk consumption per head. But this also signifies that before the period of high industrialisation considerable amounts of milk were consumed. For the retail trade this meant, bearing in mind the development of the population, a fourfold increase in demand for its services.*

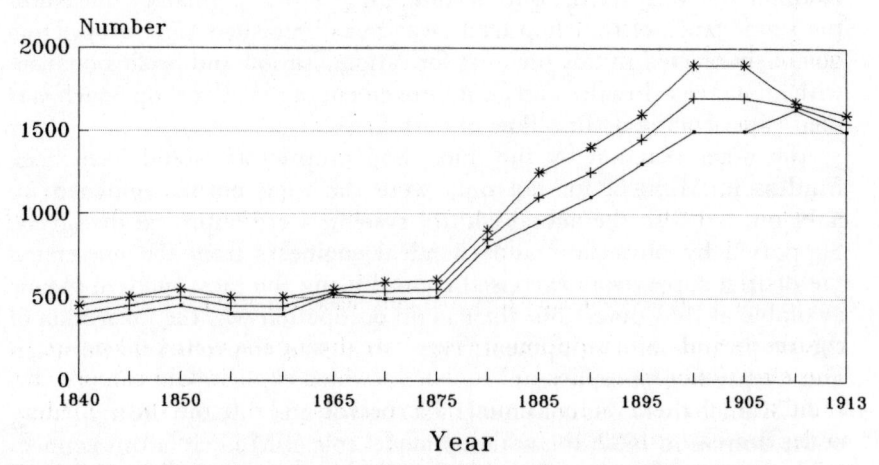

Fig. 7.3. Quantitative development of the Munich milk trade 1840–1913

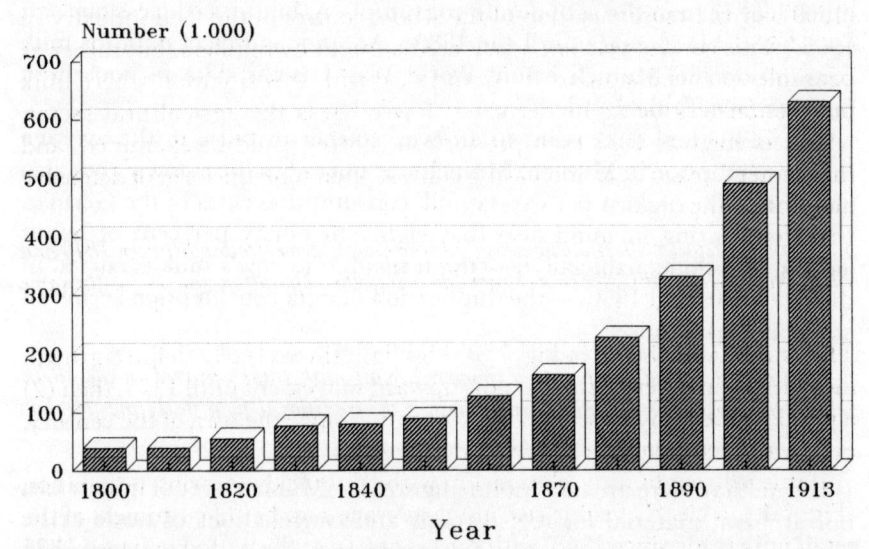

Fig. 7.4. Development of the population in Munich 1800–1913. (Source: Jahre, n.d. [1975], 131)

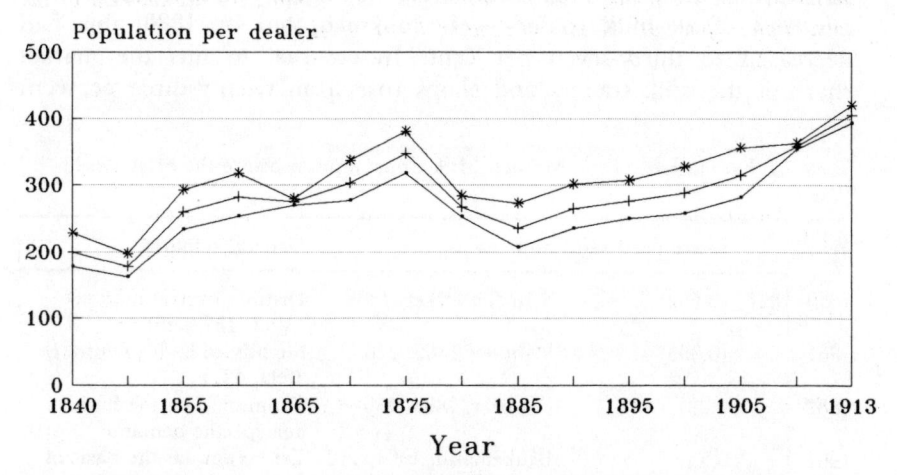

Fig. 7.5. Relation between population and milk dealers in Munich 1840–1913

Figures for milk consumption are very rare during the German Empire.[13] This situation also applies to Munich for which explicit data can only be found after the turn of the century. For periods prior to that time one has to depend on estimations (see Table 7.1).

The early consumption figures are, with respect to Munich scientists, much higher than the actual milk consumption, but nevertheless one can see a probable increase until the 1880s. Another source of data for milk consumption in Munich before World War 1 is the data in household budgets (see Table 7.2).[15]

The individual data point to an even steeper increase in the average milk consumption in Munich. Munich was not only a 'beer-town', but also had one of the highest per capita milk consumption rates in the German Empire. Bearing in mind also that eighty to ninety per cent of infant feeding was done artificially[16]—the transition to cow's milk occurred in the late 1850s and 1860s—the duplication of milk consumption seems an adequate assumption.

5. The actual radical change to a 'modern' retail milk trade occurred in the 1870s when growing competition finally undermined the traditional food stall, and the shop succeeded as the dominant medium of sale.

It is true that there are no absolute figures for Munich's retail milk trade, but archival material for registrations and cancellations of trade exists between 1868—the year when free trade had finally been achieved in Bavaria—and 1881 (see Fig. 7.6).

Until 1875 the absolute number of shops in Munich hardly varied, but this situation rapidly changed after the so-called *Gründerkrise*[17] when the actual quantitative upturn in the milk trade began. (see Fig. 7.7).

As the annual change of personnel rose to fifty per cent, the inner structure of the retail milk trade clearly changed. In 1870, seventy-two per cent of all milk traders were milkmen, but by 1880, this had decreased to thirty-seven per cent. In contrast to this the market share of the milk traders and shops rose from twenty-three per cent

Table 7.1 Estimations of the Munich Milk Consumption before the First World War.

Year	Litres per head	Sources	General remarks
1850–1867	(0,562)	Schiefferdecker, 1869, 73–4	Distinct overestimation (Voit, 1877, 25)
1861	(0,363)	Wibmer, 1862, 218.	Slightly to high (Zaubzer, 1883, 437)
1882	0,291	Zaubzer, 1883, 438–9	Estimation on the basis of age-specific demand
1902	0,359	Beukemann, 1904, 112	Estimation on the basis of railway supply
1908	0,408	Bericht, [1910], 86	ditto
1909	0,422	Reitzenstein, 1930, 46	ditto
1910	0,418	ditto	ditto
1911	0,413	ditto	ditto
1912	0,422	ditto	ditto
1913	0,424	ditto	ditto

Table 7.2 Consumption of Milk in Munich Household Budgets 1873–1915.

Year	Profession	Size of household	Litres per head per day	Time of enquiry
1873	Porter	1	Coffee w. milk	3 days
1873	Joiner Journeyman	1	Coffee w. milk	3 days
1873	Physician	1	Coffee w. milk, milk at lunch	3 days
1873	Physician	1	Coffee w. milk	3 days
1876	Inmate of home for the aged	1	0,134	7 days
1875/76	Civil Servant	3	0,167	30 days
1880	Joiner Journeyman	4	0,375	7 days
1880	Commissionaire	2	0,25	7 days
1880	Mason Journeyman	4	0,175	7 days
1880	Seamstress	1	0,214	7 days
1897	Typesetter	4	0,253	2 months
1901	Stock-keeper	2	1,036	3 months
1902	Malster	4	0,557	1 year
1902	Brewery Assistant	5	0,285	1 year
1902	Malster	5	0,296	1 year
1902	Stockroom Worker	1	0,012	1 year
1902	Malster	1	0,034	1 year
1907/08	Skilled Workers	22 Families	0,397	1 year
1908	Metal Workers	15 Families	0,362	1 year
1910	Joiner Assistant	7	0,371	1 year
1911	Joiner Assistant	7	0,388	1 year
1912	Joiner Assistant	7	0,388	1 year
1911/12	Pyrotechnician	3	0,405	1 year
1911/12	Packer Forman	3	0,326	1 year
1912	Artist Painter	4	0,253	1 year
1913	Artist Painter	4	0,263	1 year
1914	Artist Painter	4	0,271	1 year
1913	Civil Servant	7	0,429	1 year
1913	Civil Servant	7	0,435	1 year
1914	Type setter	5	0,353	1 year
1915	All	4616 Families	0,412	1 month

to sixty-two per cent.[18] Thus the shop as the main medium of sale was finally fully accepted by the dealers, while the producers were later on consistently pushed out of the retail trade. The percentage of shops with at least ten years continued existence are clear evidence of the change from the food stall with its traditional methods and orientation, to a modern competitive food sector (see Fig. 7.8).

This figure clearly shows the small personnel mobility of the Munich retail trade until the middle of the 1870s. Changes which had occurred

78 *Milk and milk products*

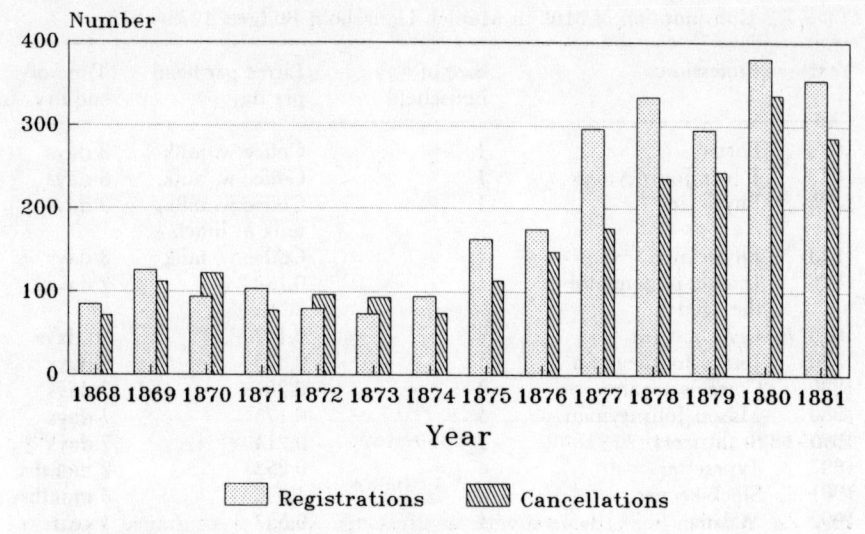

Fig. 7.6. Registrations and cancellations of trade in the retail milk trade 1868–1881. (Source: StA München Gewerbeamt 104/3).

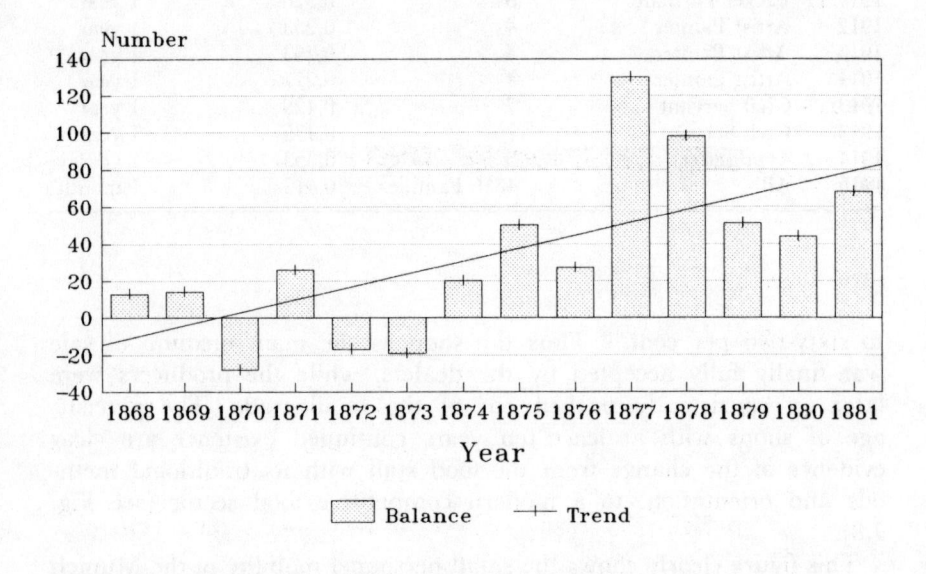

Fig. 7.7. Fluctuations in the Munich milk trade 1868–1881. (Source: StA München Gewerbeamt 104/3)

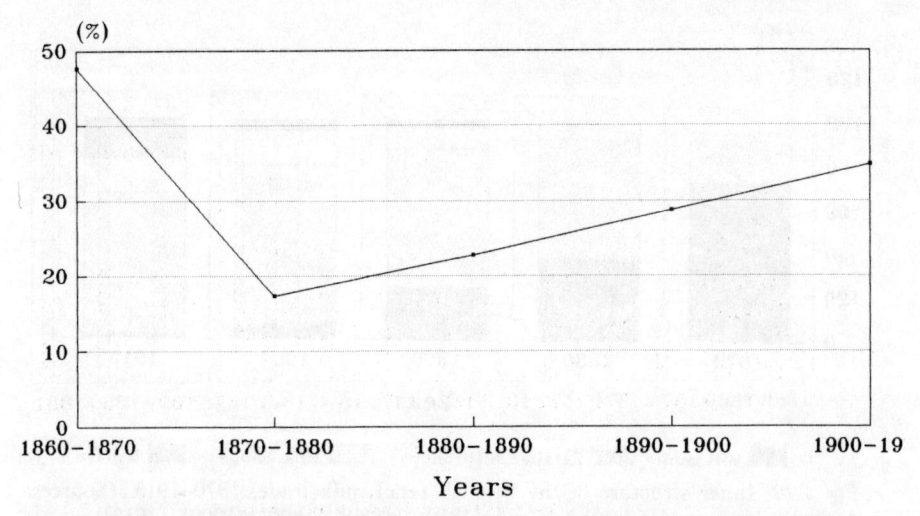

Fig. 7.8. Munich milk dealers with at least ten years of continued existence 1860–1910. (Source: Author's analysis of Adressbuch, n.d. [1860]. Adreßbuch, n.d. [1870], [1880], [1890], [1900], [1913])

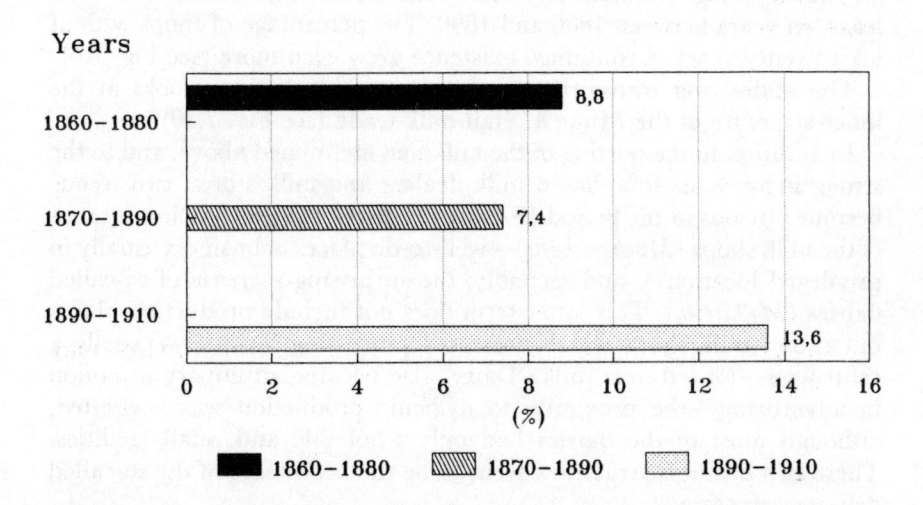

Fig. 7.9. Munich milk dealers with at least twenty years of continued existence 1860–1910. (Sources: Author's analysis of Adressbuch, n.d. [1860]. Adreßbuch, n.d. [1870], [1880], [1890], [1900], [1913])

80 *Milk and milk products*

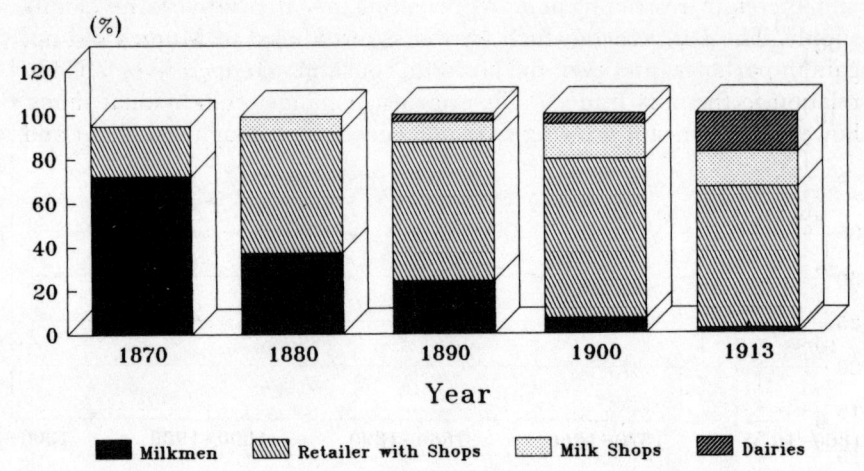

Fig. 7.10. Inner structure of the Munich retail milk trade, 1870–1913. (Sources: Author's analysis of Adreßbuch, n.d. [1870], [1880], [1890], [1900], [1913])

were mostly due to the addition of new elements—the milk storehouses, for example—while the old trade mechanisms survived.

6. *After the period of transition 1875–1885, the Munich milk trade consolidated noticeably and, on average, milk retailers drew close to the norms of a secure middle-class existence. Milk prices, buying on credit, and increasing shop rents, became the decisive economic data which determined future development. An increase in sales was the rational answer to the increase in fixed costs.*

As shown in Fig. 7.8 more and more retailers managed to survive for at least ten years between 1880 and 1890. The percentage of shops with at least twenty years of continued existence grew even more (see Fig. 7.9).

The underlying transformation becomes evident if one looks at the inner structure of the Munich retail milk trade (see Fig. 7.10).

In addition to the ousting of the milkmen mentioned above, and to the strong increase in shop-based milk dealers and milk stores, two trends become obvious in the period 1870–1913: firstly, the growing importance of the milk shops (*Milchgeschäfte*)—well-to-do places of business usually in privileged locations—and secondly, the surprising increase of so-called dairies (*Molkereien*). This latter term does not include production plants but rather milk shops which possessed processing facilities—usually a centrifuge—for left-over milk. 'Dairy' also became an important notion in advertising—the proximity to hygienic production was suggested, although most of the dairies had only wholesale and retail facilities. These two business structures seem to be the precursors of the so-called delicatessen store.

They were fairly efficient enterprises capable of controlling the market

The retail milk trade in transition 81

and, therefore, restricting or even preventing new alternative forms of milk supply. The *Bassinwagen* which were first introduced in Munich did not gain importance, and even the powerful consumer co-operatives failed in relation to the milk trade.[19] The growing demands on individual shops, however, also meant growing risks. The customers bought on credit and

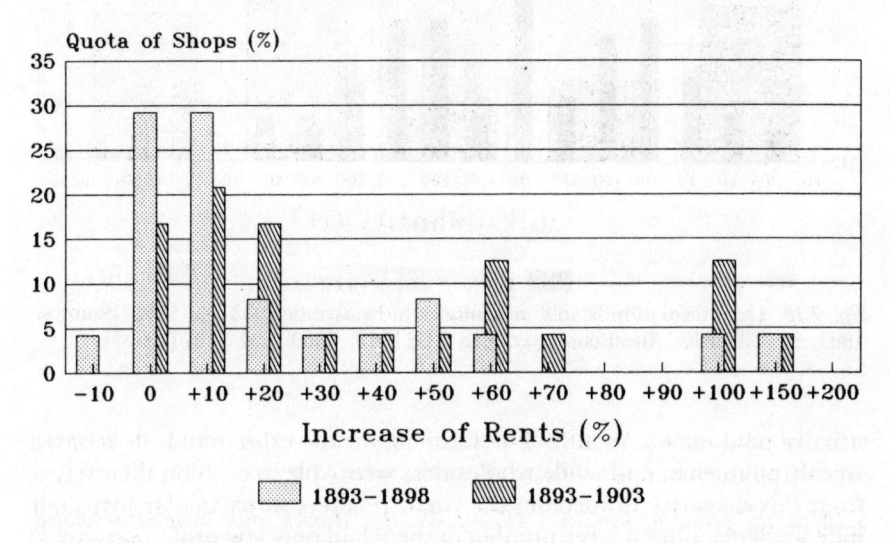

Fig. 7.11. Development of milk shop rents in Munich-Ludwigstraße 1893–1903. (Source: StA München Bezirksinspektionen 505)

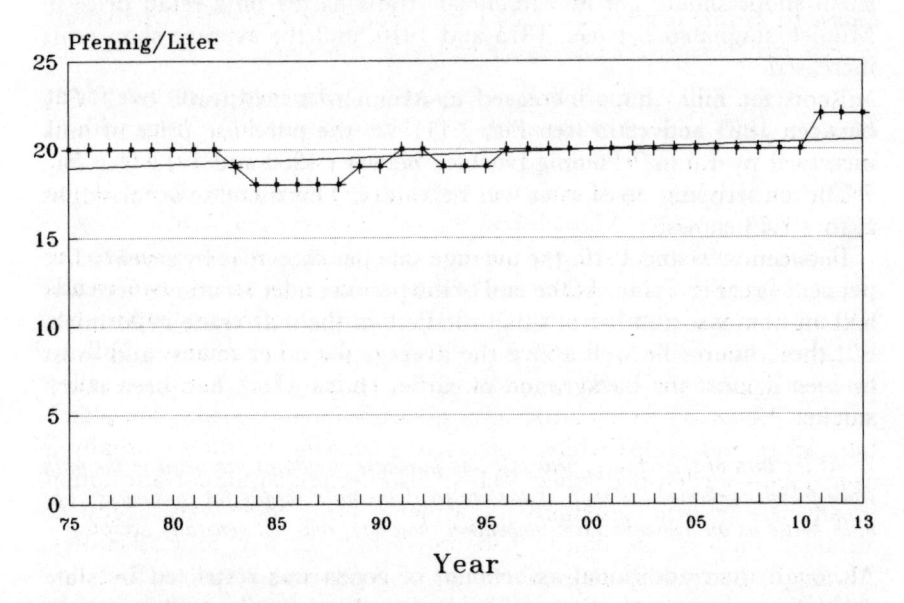

Fig. 7.12. Retail milk prices in Munich 1875–1913. (Sources: Milchwirtschaft, 1910, 208–209; Morgenroth, 1914, 295)

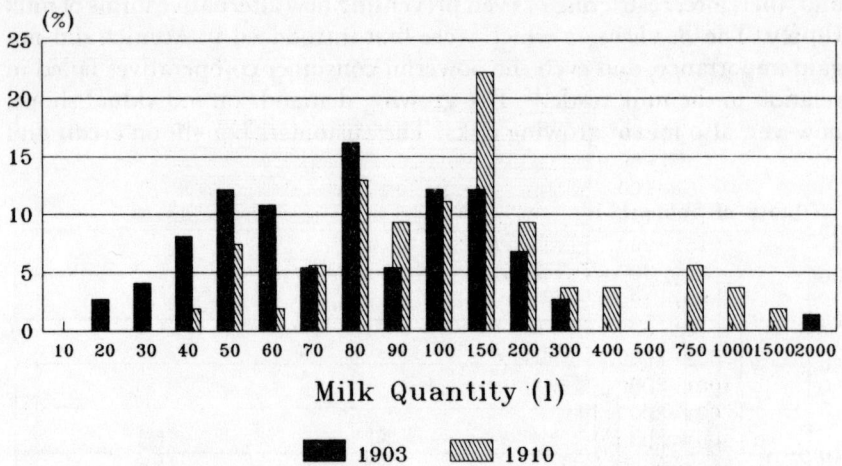

Fig. 7.13. Quantities of milk sold in Munich-Ludwigstraße 1903 and 1910. (Sources: 1903: StA München Bezirksinspektionen 505; 1910: Fiack, n.d. [1910], 6–7).

usually paid once a month. The farmers, on the other hand, demanded weekly payments, and while wholesalers were able to cushion themselves from this disparity in income, the small retailers in particular lost their independence since a large number of them had only low profit margins,[20] unlike the prospering milk shops and dairies. It was inevitable that these small shops should get into financial straits as the milk retail price in Munich stagnated between 1875 and 1910, and the average shop rents increased.

Rents for milk shops increased in Munich-Ludwigstraße by 27.6% between 1893 and 1903 (see Fig. 7.11). As the purchase price of milk increased by 1.5 to 2 Pfennig per litre between 1900 and 1913 (see Fig. 7.12), the stepping up of sales was necessary. This increase occurred, as Figure 7.13 shows:

Between 1903 and 1910, the average sale per shop rose by seventy-five per cent (see Fig. 7.14). At the end of the period under scrutiny, there was still an immense number of small retailers in the milk trade in Munich, but these figures lie well above the average for other towns, and must be seen against the background of earlier shops which had been much smaller.[21]

7. *At the turn of the century, new sale and purchase standards prevailed in the milk trade which confirmed the shop as the centre of the milk supply for consumers. The milk bottle as an hygienic basic innovation, however, was not generally accepted.*

Although their additional assortment of goods was restricted by state regulations, most milk shops sold other products besides milk.

According to Figure 7.15, milk shops were actually both milk and bread

Milk Quantity (l)

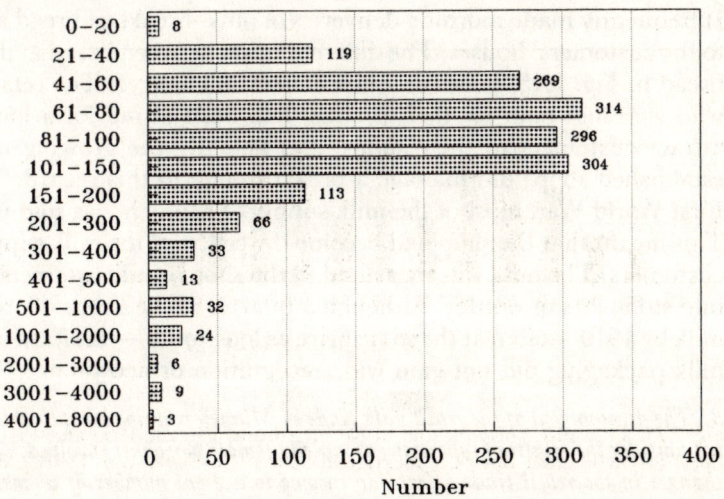

Range	Number
0–20	8
21–40	119
41–60	269
61–80	314
81–100	296
101–150	304
151–200	113
201–300	66
301–400	33
401–500	13
501–1000	32
1001–2000	24
2001–3000	6
3001–4000	9
4001–8000	3

Fig. 7.14. Quantities sold in the Munich retail milk trade, 10.10.1910. (Source: Flack, n.d. [1910], 6–7)

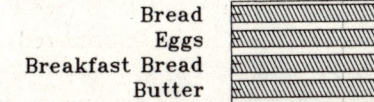

Articles

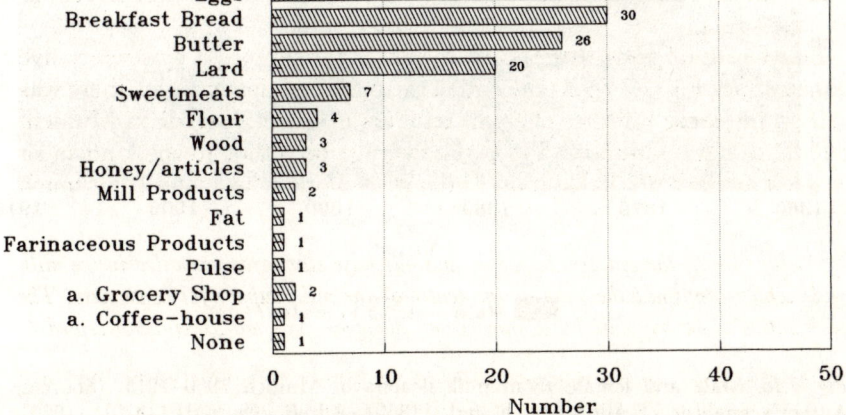

Article	Number
Bread	37
Eggs	37
Breakfast Bread	30
Butter	26
Lard	20
Sweetmeats	7
Flour	4
Wood	3
Honey/articles	3
Mill Products	2
Fat	1
Farinaceous Products	1
Pulse	1
a. Grocery Shop	2
a. Coffee-house	1
None	1

Fig. 7.15. Goods' assortment of milk shops in Munich-Ludwigstraße 1903. (Source: StA München Bezirksinspektionen 505)

84 *Milk and milk products*

shops, which also sold dairy products; shops which sold milk exclusively were unusual in Munich. The retail trade also provided service functions. It frequently made morning deliveries of milk, breakfast bread and butter to the customers' houses. The differentiation between bread and breakfast bread in Fig. 7.15, however, hints at certain changes. The retail dealers, who were not paid for their service, had been trying for a long time to attract customers to their shops, and actually the growing number of established shops did manage a breakthrough in this regard. Before the First World War, most of the milk supply of Munich was sold in shops.[22] This meant that the shop had become the centre of the milk supply for the customers. The milk was measured in the shops, and customers poured it into stone jars or bottles. Although a quarter of the shops offered bottled milk by 1910—often at the same price as loose milk—this form of hygienic milk packaging did not gain wide recognition or acceptance.[23]

8. The development of the retail milk trade in Munich remains inexplicable if one does not consider the relative proportions of male to female personnel involved. Although the changes in the retail trade offered an income to a great number of women, the main beneficiaries of the changes were men. The milk trade as a whole was characterised by an ever-widening gap between the increasing profits of men and the increasing self-exploitation of women (see Fig. 7.16).

The retail milk trade in Munich became increasingly female in character as the percentage of women in the entire sector grew rapidly. But while the number of men in the milk trade dropped, their position remained dominant. The men usually ran well-established shops with good chances

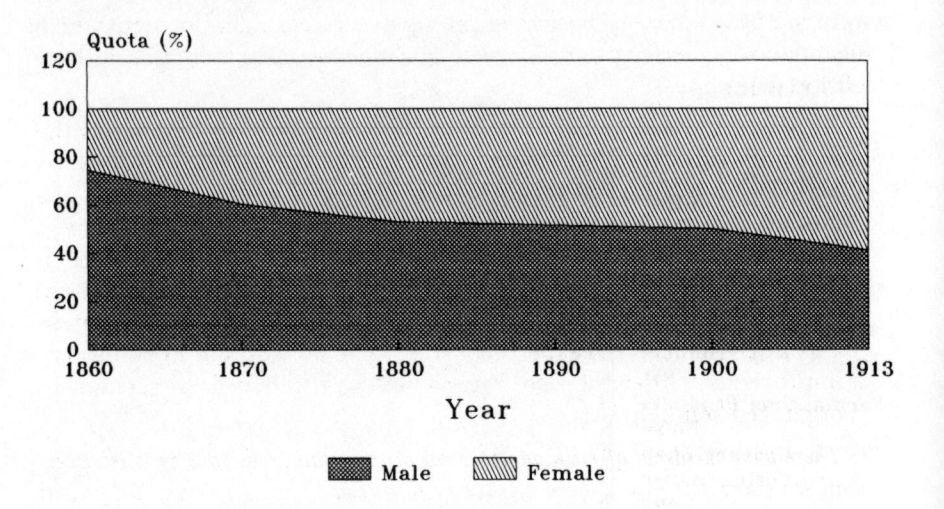

Fig. 7.16. Male and female retail milk dealers in Munich 1860–1913. (Sources: Author's analysis of Addressbuch, n.d. [1860]. Adreßbuch, n.d. [1870], [1880], [1890], [1900], [1913])

The retail milk trade in transition 85

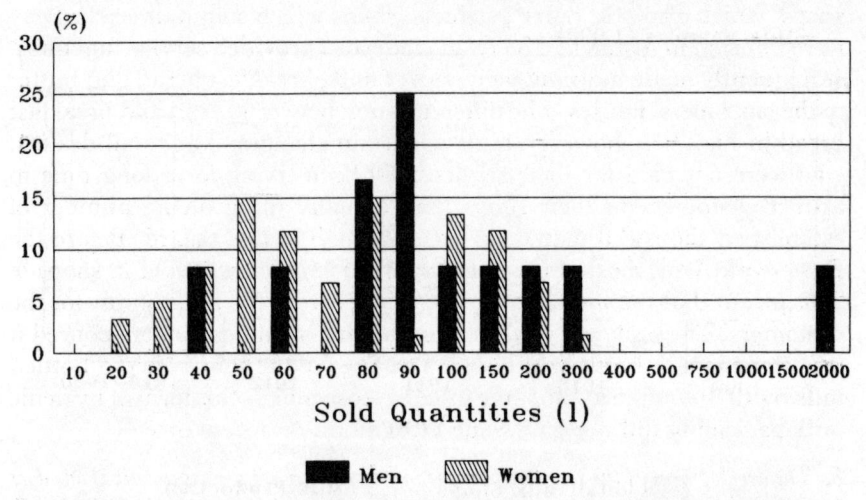

Fig. 7.17. Sales volume according to the sex of the retailers in Munich-Ludwigstraße 1903. (Source: Author's analysis of StA München Bezirksinspektionen 505)

of making profits, while women managed meagre part-time businesses, mostly as dependent functionaries of the wholesale dealers.

Proof of this situation is the larger quantities of milk sold in shops run by men. In the district of Munich-Ludwigstraße, these were three times as high, i.e. 270 litres sold by men compared to 82 sold by women (see Fig. 7.17). Men also owned the majority of the new, well-to-do dairies. Only 19 of 113 dairies for the year 1913, (i.e. about seventeen per cent) were run by women.[24] The average numbers conceal a development which meant above-average profits for men. For many women there was nothing left for them but:

> An unhappy, dreary life, illness and premature death because of the incessant long working hours (from five in the morning to nine at night), an inferior offspring which is denied the blessings of the mother's breast (because the women has to take up her business again four or five days after the delivery), a wretched existence in damp cellar dwellings; these are some of the worst characteristics of this kind of working life. And the worst is that years of destitution and oppression have made them dull, that they do not hope for any improvement, that they are impervious to the beneficial thought of organized self-help.[25]

9. *There were incidents of milk adulteration in the retail trade, but on a relatively small scale and this also showed a tendency to decline. The agricultural sector was much more problematic in terms of an hygienic milk supply.*

The milk co-operatives had a positive attitude towards the considerable controls in Munich, which led to a milk supply that met

86 *Milk and milk products*

[Bar chart: Milk Samples (1.000) for years 1909, 1910, 1911, 1912, 1913–1920, showing Retail Milk Shops and Milk Production]

Retail Milk Shops   Milk Production

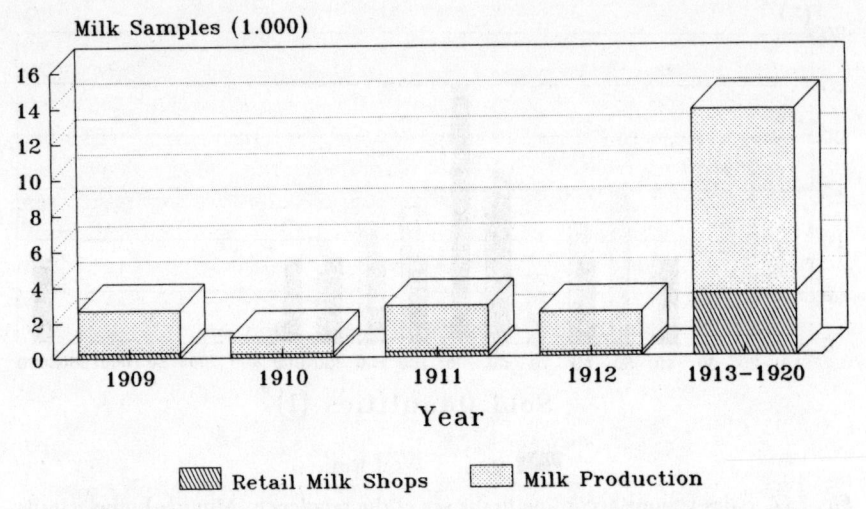

Fig. 7.18. Adulterated milk samples in retail trade and milk production 1909–1920. (Sources: Bericht, n.d. [1910–1913]. Verwaltungsbericht, n.d. [1920])

contemporary standards. Nevertheless, more than 1,000 cases of milk adulteration per annum were reported at the beginning of the twentieth century. But it is most instructive to see where these adulterations occurred.

According to Fig. 7.18, there had been one milk adulteration per ten to fifteen milk shops per year, which was hardly a very serious situation. It was in the area of milk production, however, as Fig. 7.19 hints, that the real problem lay; the rate of milk adulteration here was four to ten times greater than in the retail trade.

10. The contemporary debate about an hygienic milk trade was by no means a discourse about real conditions. Being only insufficiently based on facts, it has to be understood, on the one hand, as a discourse by doctors and hygienists aimed at establishing themselves professionally, and on the other, as the conspiratorial talk of men about women.

Contemporary scholars were not aware of this development of the milk trade. Their comments on the milk supply portrayed the retail milk trade as being very dubious indeed. Neither did they think in terms of the development the retail trade had gone through in the past, or consider the perspective of the consumer who had to rely on the decentralised distribution of milk. Most scholars thought only in absolute terms in relation to their technical and scientific norms and, therefore, were quick to call for 'elimination' and 'eradication' of the retail trade. To understand their restricted perspective, one has to consider also the scholars' personal positions; as persons with adequate positions, they were the people most suited to meet the standards they themselves had set.

At the same time it is surprising that the discussion among exclusively male scholarly elites concentrated on the smallest shops in the milk trade. These shops, usually run by women, were described in a mode of language which implies repressed fears of women. Furthermore, the rigidity of the solutions implies hierarchical treatment of the other sex. Clearly one has to follow up such hints carefully, but it would seem that the picture of the retail milk trade painted by contemporary commentators is potentially very different from the one which I see. Thus, while the views of contemporary commentators must be carefully considered, nevertheless it is now obvious from my research that further investigation of this topic requires that, in addition to applying the usual methods of source criticism to the documentary sources, we

Fig. 7.19. *Die panschenden Bauern*[26]

'Also, daß Du's weißt, der Apotheka hat g'esagt, auf an' Liter von dera Medizin für die Kuah kommt e' viertel liter wasser!'—'Guat, i' versteh' schon ... des is ja g'rad' wia bei da milk!' ('So that you know, the pharmacist said to add a quarter of a litre of water to one litre of medicine for the cow!' 'Well, I understand ... it is just the same as with the milk!')

88 Milk and milk products

must also recognise in them something of a contemporary expression of *zeitgeist*.[27]

NOTES AND REFERENCES

1 In contrast to Great Britain the history of retailing in Germany is virtually unexplored, but I hope to give some first impressions in this regard in my forthcoming doctoral thesis: '*Kleinhandel in Kaiserreich. Fallstudien zu Hamburg und München*'.
2 This lecture is a condensed version of the author's larger unpublished text '*Milchkleinhandel im Wandel. Eine Fallstudie zu München 1840–1913*', containing comprehensive lists of sources and literature.
3 These and following are from *StA München Gewerbeamt 5322* . . .
4 Cf. G. Bauschinger, *Das Verhältnis von städtischer Selbstverwaltung und königlicher Polizei in München im 19. Jahrhundert*, Jur. Diss., München, 1968; M. Block, *Die Verfassung der Stadt München von 1818–1819*, Jur. Diss., München, 1967.
5 For a general survey see Pappe, 1975.
6 Feser, 1878, 7: 'In München, wo schon seit 30 Jahren eine geordnete, bisher von den städtischen Thierärzten gut geleitete Milchcontrole existirt, sind bedeutendere Milchfälschungen verhältnismässig selten'. (Deliberate milk adulteration is relatively seldom in Munich where well-regulated milk control, previously undertaken by veterinary surgeons, has existed for 30 years.) For a contrary view, cf. Hauner, 1858, 132–133.
7 See especially: 'Schreiben der Königlichen Regierung von Oberbayern an den Stadtmagistrat München' v. 26.1.1861 in *StA München Gewerbeamt 5322* . . .
8 'Milchverkauf- und Beschauordnung', 12.8.1862 in *Adreßbuch*, n.d. [1879], 121–122.
9 E.g. 'Instruction . . .', 1877, 172–174.
10 'Ortspolizeiliche Vorschrift über den Verkehr mit Nahrungs- und Genußmitteln v. 12.1.1892, in *Adreßbuch*, n.d. [1894], 123–126; 'Bayern . . .', 1901; 'Ortspolizeiliche Vorschriften v. 5.10.1906, den Verkauf von Nahrungs- und Genußmitteln betreffend', in *Adreßbuch*, n.d. [1907], 209–210. Clevisch, 1909, 22 tells something about the national model role of Munich.
11 See Sendtner, 1901, 1109.
12 The best surveys of milk production are *Die Milchwirtschaft* . . ., 1910, and Arnold, 1911.
13 Data on the basis of *Adressbuch* . . ., n.d. [1860]; *Adreßbuch* . . ., n.d. [1870], [1880], [1890], [1900], and [1913]; *StA München Gewerbeamt* . . ., 106/1, *Ibid.* 104/3; '*Bericht* . . .', 1889; 'Die Landwirthschaft . . .' 1890; Creuzbauer, 1903; '*Bericht* . . .', n.d. [1909–1913]; *Gewerbestatistik* . . ., 1886; *Gewerbe-Statistik* . . . 1898; *Gewerbliche Betriebsstatistik* . . ., 1910; Fiack, n.d. [1910].
14 Teuteberg, 1976 contains only data for all milk products. *Idem.* Teuteberg, 1981.
15 Forster, 1873, 387–390; Forster, 1877, 192, 214; Dehn, 1880, 852–855; Abelsdorff, 1900, Anhang; *Erhebungen* . . ., 1902, 24; *Erhebungen* . . ., 1903, 32–33; Conrad, 1909, 54; *Haushaltungsrechnungen* . . ., 1909, 82; 'Warum müssen wir Arbeiter . . .?', 1912; 'Ein Arbeiter-Haushaltungsbudget . . .' 1913; Krziza, 1915, 244; Jentzsch, 1920, 325; Patschoky, 1918, 605, 613; Welker, 1916, 95.
16 Escherich, 1887.
17 The *Gründerkrise*, 1873, resulted from the economic speculation victory foundation of the German Empire, 1870–1871 and began a period of economic depression which lasted until 1879 at least. A quantitative upturn in the Munich retail milk trade was possible after the so-called *Gründerkrise* since: (1) it was a crisis

affecting primarily heavy industry and the financial sector, (2) the rents of many shops were quite cheap, and (3) it was quite attractive to invest in the consumer sector since it was less sensitive to the depression. During the depression began the real spatial growth of the town and because of this many shops were established in less-populated areas of the town. Cf. R.H. Tilly, *Vom Zollverein zum Industriestaat. Die wirtschaftlich-soziale Entwicklung Deutschlands 1834 bis 1914*, München, 1990, 78–83.

18 *Adreßbuch*..., n.d. [1871]; n.d. [1881]. These data are based on the author's own analysis of *Adreßbuch*..., n.d. [1871] – n.d. [1881]. As the Munich Directories do not contain any information about the retail milk trade in their trade and business chapters, it was necessary for the writer to check the names, professions and addresses of nearly 600,000 people on the alphabetical list of inhabitants in order to arrive at the presented data.

19 Information about the *Bassinwagen* in *Ein Wegweiser*..., 1912; Erlbeck, 1911, 64; Dallmayr, 1912; *Bericht*..., n.d. [1907]. Sources for the milk trade of Munich consumer co-operatives are particularly *Der Konsumverein*... 1906, 944, 1147, Beck, n.d. [1921], 105–106, 116–117 and *Konsum-Verein*..., 1926, 39.

20 E.g. Creuzbauer, 1903.

21 The Hamburg chemist Dunbar estimated 60–80 litres sales per day in 1903 in the German Empire (Prölls, 1904, 530). Munich showed 120 litres in 1910 (without grocery milk).

22 Clevisch, 1909, 22–23.

23 Fiack, n.d. [1910], 8.

24 These data are based on author's own analysis of *Adreßbuch*..., n.d. [1914]. Cf. also note 16 above.

25 Dallmayr, 1912, 88.

26 ('Adulterating farmers') from 'Schnell begriffen', *Fliegende Blätter*, 139, 1913, Nr. 3546, 10.

27 Because of restrictions of space many additional aspects of the Munich milk trade could not be dealt with in this paper: for example, the associations in the retail trade, the integration of the milk trade in discount savings societies, the interaction between the retail and the wholesale trade, the question of milk wars, changes in production, and so on. It is likely that future research into these problems will qualify some of my statements in this paper.

SOURCES AND LITERATURE

Primary Sources

StA München Bezirks-Inspektionen 505: Betreff: Statistische und sonstige Erhebungen (1881–1920).

StA München Gewerbeamt 104/3: Act des Magistrats der königl. Haupt- & Residenzstadt München. Gewerbestatistik 1870–1887.

StA München Gewerbeamt 106/1: Vergleichende Zusammenstellung der Zahl der Gewerbe u. freien Erwerbsarten in München in den Jahren 1846, 1861 u. 1867 nebst Auszug hieraus.

StA München Gewerbeamt 5322: General-Acten des Magistrats der Königl. Haupt- und Residenzstadt München. Betreff Milchhandel in specie Milchniederlagen (1838–1861).

Secondary Sources

W. Abelsdorff, *Beiträge zur Sozialstatistik der Deutschen Buchdrucker*, Tübingen/Leipzig, 1900.

Adreßbuch für München und Umgebung 1914, München, n.d. [1913].

Adreßbuch für München und Umgebung 1908, München, n.d. [1907].
Adreßbuch von München für das Jahr 1901, hg. v. d. kgl. Polizei-Direktion, München, n.d. [1900].
Adreßbuch von München für das Jahr 1895, hg. v. d. kgl. Polizei-Direktion, München, n.d. [1894].
Adreßbuch von München für das Jahr 1891, hg. v. d. kgl. Polizei-Direktion, München, n.d. [1890].
Adreßbuch von München für das Jahr 1881, hg. v. d. kgl. Polizei-Direktion, München, n.d. [1880].
Adreßbuch von München für das Jahr 1880, hg. v. d. kgl. Polizei-Direktion, München, n.d. [1879].
Adreßbuch von München für das Jahr 1871. Nach amtlichen Quellen bearb. v. M. Siebert, München, n.d. [1870].
Adressbuch von München für das Jahr 1861. Im Auftrage der Königlichen Polizeidirektion München aus amtlichen Quellen bearb. v. M. Siebert, München, n.d. [1860].
'Warum müssen wir Arbeiter uns der Sozialdemokratie anschließen?', Münchener Post, (1912), Nr. 8 v. 12.01.
'Ein Arbeiter-Haushaltungsbudget', Münchener Post, (1913), Nr. 4 v. 05.01.
P. Arnold, 'Zur Frage der Milchversorgung der Städte', Jahrbücher für Nationalökonomie und Statistik, **96** (1911), 585–642.
'Bayern. München. Ortspolizeiliche Vorschriften über den Verkauf von Kindermilch. Vom 29. Dezember 1899', Zeitschrift für Untersuchung der Nahrungs- und Genußmittel, **4** (1901), 45.
H. Beck, Die Konsumvereine Münchens, RStWiss. Diss., Werzburg, n.d. [1921].
'Schnell begriffen', Fliegende Blätte, **139** (1913), Nr. 3546, 10.
'Bericht über die Ergebnisse der berufsstatistischen Erhebung vom 5. Juni 1882 in München', in: Mittheilungen des Statistischen Bureaus der Stadt München, Bd. 7, München, 1889, 1–338.
Bericht über den Stand der Gemeinde-Angelegenheiten der Kgl. Haupt- und Residenzstadt München für das Jahr 1912, Verwaltungsbericht, München, n.d. [1913].
Bericht über den Stand der Gemeinde-Angelegenheiten der Kgl. Haupt- und Residenzstadt München für das Jahr 1911, T. I: Verwaltungsbericht, München, n.d. [1912].
Bericht über den Stand der Gemeinde-Angelegenheiten der Kgl. Haupt- und Residenzstadt München für das Jahr 1910, T. 1: Verwaltungsbericht, München, n.d. [1911].
Bericht über den Stand der Gemeinde-Angelegenheiten der Kgl. Haupt- und Residenzstadt München für das Jahr 1909, T. I: Verwaltungsbericht, München, n.d. [1910].
Bericht über den Stand der Gemeinde-Angelegenheiten der Kgl. Haupt- und Residenzstadt München für das Jahr 1908, T. 1: Verwaltungsbericht, München, n.d. [1909].
Bericht über den Stand der Gemeinde-Angelegenheiten der Kgl. Haupt- und Residenzstadt München für das Jahr 1906, T. 1: Verwaltungsbericht, München, n.d. [1907].
Gewerbliche Betriebsstatistik, Abt. V: Großstädte, hg. v. Kaiserlichen Statistischen Amte, Berlin, 1910. (Statistik des Deutschen Reichs, Bd. 217).
W. Beukemann, 'Der Milchverbrauch der Städte. Sein Zusammenhang mit den Landesverhältnissen und dem Wohlstande der Bewohner', in: Bericht über die Allgemeine Ausstellung für hygienische Milchversorgung im Mai 1903 zu Hamburg, hg. v. d. Deutschen Milchwirtschaftlichen Verein, Hamburg, 1904, 95–130.
A. Clevisch, Die Versorgung der Städte mit Milch, Hannover, 1909.
E. Conrad, Lebensführung von 22 Arbeiterfamilien Münchens. Im Auftrage des Statistischen Amtes der Stadt München dargestellt, München, 1909. (Einzelschriften des Statistischen Amtes der Stadt München, No. 8).
A. Creuzbauer, Die Versorgung Münchens mit Lebensmitteln. Eine volkswirtschaftliche Studie, München, 1903.
A. Dallmayr, Milchversorgung und Milchkriege der Stadt München. Ein Beitrag zur Lösung der 'Milchfrage', München, 1912.

P. Dehn, 'Deutsche Haushaltungsbudgets. III. Bayerische Budgets', *Annalen des Deutschen Reichs*, (1880), 843–855.
Erhebungen der königlich Bayerischen Fabriken- und Gewerbe-Inspektoren über das Bierbrauergewerbe (Beilagenheft zu den Jahresberichten für 1902.), München, 1903.
Erhebungen der königlich Bayerischen Fabriken- und Gewerbe-Inspektoren über das Müllergewerbe (Beilagenheft zu den Jahresberichten für 1901), München, 1902.
A.R. Erlbeck, 'Städtische Milchversorgung', *Soziale Revue*, **11** (1911), 56–69.
H. Escherich, 'Die Ursachen und die Folgen des Nichtstillens bei der Bevölkerung Münchens', *Münchener Medizinische Wochenschrift*, **34** (1887), 233–235, 256–259.
J. Feser, *Die Polizeiliche Controle der Markt-Milch. Zwei Vorträge*, Leipzig, 1878.
Fiack: *Milchversorgung der Stadt München am 10. Oktober 1910.* (Ergebnisse einer besonderen Erhebung.). Sonderabdruck aus Mitteilungen des Statistischen Amtes der Stadt München, Bd. XXIII, s.l., n.d. [München 1910].
J. Forster, 'Ueber die Kost in Armen- und Arbeitshäusern', in C. Voit (Hg.), *Untersuchung der Kost in einigen öffentlichen Anstalten. Für Aerzte und Verwaltungsbeamte zusammengestellt*, München, 1877, 186–215.
J. Forster, 'Beiträge zur Ernährungsfrage', *Zeitschrift für Biologie*, 9 (1873), 381–410.
Gewerbestatistik nach der allgemeinen Berufszählung vom 5. Juni 1882, 1. Gewerbestatistik des Reichs und der Großstädte, Th. 2: Gewerbestatistik der Großstädte, hg. v. Kaiserlichen Statistischen Amt, Berlin, 1886. (Statistik des Deutschen Reichs NF, Bd. 6, Th. 2).
Gewerbe-Statistik der Großstädte, bearb. im Kaiserlichen Statistischen Amt, Berlin, 1898. (Statistik des Deutschen Reichs NF, Bd. 116).
Hauner: 'Bericht über das eilfte (sic!) Jahr des mit dem Kinderhospitale zu München verbundenen Ambulatoriums', *Journal für Kinderkrankheiten*, **31**, (1858), 121–155.
320 Haushaltungsrechnungen von Metallarbeitern, bearb. u. hg. v. Vorstand des Deutschen Metallarbeiter-Verbandes, Stuttgart, 1909.
'Instruction für die Markt- und Bezirks-Inspectoren zur Vornahme der Victualien-Beschau in München', *Correspondenz-Blatt des Niederrheinischen Vereins für öffentliche Gesundheitspflege*, **6**, (1877), 166–177.
1875–1975. 100 Jahre Städtestatistik in München, München, n.d. [1975].
'Münchener Jahresübersichten für 1906', in: *Mitteilungen des Statistischen Amtes der Stadt München*, Bd. 21, München, 1907, H. 2, 1*–100.*
'Münchener Jahresübersichten für 1903', in: *Mitteilungen des Statistischen Amtes der Stadt München*, Bd. 18, München, 1904, H. 4, 1*–96*.
'Münchener Jahresübersichten für 1895', in: *Mittheilungen des Statistischen Amtes der Stadt München*, Bd. 14, München, 1896, 37*–72.*
W.H. Jentzsch, 'Das Wirtschaftsbuch eines Kunstmalers', *Allgemeines Statistisches Archiv*, **12**, (1920), 316–329.
Konsum-Verein Sendling München 1886–1926. Jubiläums-Schrift anlässlich des 40jährigen Bestehens des Vereins und Jahres-Bericht über das 40. Geschäftsjahr vom 1. Juli 1925 bis 30. Juni 1926, München, 1926.
'Der Konsumverein Sendling-München (e.G.m.b.H.)', *Konsumgenossenschaftliche Rundschau*, **3**, (1906), 943–944.
'Der Konsumverein Sendling-München', *Konsumgenossenschaftliche Rundschau*, **3**, (1906), 1147.
A. Krziza (Bearb.), *259 deutsche Haushaltungsbücher geführt von Abonnenten der Zeitschrift 'Nach Feierabend' in den Jahren 1911–1913*, Leipzig, 1915.
Die Landwirthschaft in Bayern. Denkschrift nach amtlichen Quellen bearb., München, 1890.
'Der Münchener Marktverkehr in den Jahren 1890 und 1892 mit Rückblicken auf die Vorjahre', in: *Mittheilungen des Statistischen Amts der Stadt München*, Bd. 11, München, 1893, 301–391.
Die Milchwirtschaft in Bayern, München, 1910. (Beiträge zur Statistik des Königreichs Bayern, H. 78).

W. Morgenroth, 'Die Kosten des Münchener Arbeiterhaushalts in ihrer neueren Entwicklung', in: F. Eulenburg (Hg.), *Kosten der Lebenshaltung in deutschen Großstädten, T. 2: West- und Süddeutschland, München/Leipzig*, 1914. (Schriften des Vereins für Socialpolitik, Bd. 145, T. 2), 269–305.

F. Patschoky, 'Untersuchungen über die Lebenshaltung bayerischer Familien während des Kriegs (II)', *Zeitschrift des K. Bayerischen Statistischen Landesamts*, **50** (1918), 592–626.

Prölls: 'Die Milchversorgung unserer Grossstädte unter Anlehnung an die Hamburger Milchausstellung 1903', *Deutsche Vierteljahrsschrift für öffentliche Gesundheitspflege*, **36** (1904), 508–534.

E. Frhr. v. Reitzenstein, *Die Milchversorgung der Stadt München, ein Beitrag zur oberbayerischen Milchwirtschaft*, Techn. Diss. TU München, Kempten, 1930.

W. Schiefferdecker, *Ueber die Ernährung der Bewohner Königsberg's und anderer grosser Städte. Ein Vortrag, Königsberg, 1869.*

R. Sendtner, 'Ueber die Bedeutung der ambulanten Thätigkeit bei der Ausübung der Lebensmittelkontrolle', *Zeitschrift für Untersuchung der Nahrungs-und Genußmittel*, **4** (1901), 1106–1115.

Verwaltungsbericht der Landeshauptstadt München 1913–1920 (1. Januar 1913 bis 31. März 1920), bearb. v. Statistischen Amte der Stadt München, München, n.d. [1920].

C. Voit, 'Ueber die Kost in den Volksküchen', in Ders. (Hg.), *Untersuchung der Kost in einigen öffentlichen Anstalten. Für Aerzte und Verwaltungsbeamte zusammengestellt*, München, 1877, 14–65.

'Ein Wegweiser für hygienischen Milchausschank', *Concordia*, **19** (1912), 271.

G. Welker, *Die Münchener Erhebung über den Lebensmittelverbrauch im Februar 1915. Eine statistische Studie*, München/Berlin/Leipzig, 1916.

C. Wibmer, *Medizinische Topographie der k. Haupt-und Residenzstadt München*, H. 2, München, 1862.

Zaubzer: 'Zum Milchconsum der Stadt München im Jahre 1882', *Annalen der städtischen allgemeinen Krankenhäuser zu München*, **7** (1883), 437–443.

Literature

O. Pappe, *Zur Geschichte der Lebensmittelüberwachung im Königreich Bayern (1806–1918)*, Chem. Diss., Marburg, 1975.

H.J. Teuteberg, 'The Beginnings of the Modern Milk Age in Germany', in: A. Fenton/T.M. Owen (Eds.), *Food in Perspective. Proceedings of the Third International Conference on Ethnological Food Research, Cardiff, Wales, 1977*, Edinburgh, 1981, 283–311.

H.J. Teuteberg, 'Der Verzehr von Nahrungsmitteln in Deutschland pro Kopf und Jahr seit Beginn der Industrialisierung (1850–1975). Versuch einer quantitativen Langzeitanalyse', *Archiv für Sozialgeschichte*, **19** (1976), 331–388.

SELECT BIBLIOGRAPHY

F. Anderegg, *Allgemeine Geschichte der Milchwirtschaft (vom Altertum bis zur Gegenwart)*, Zürich, 1894.

P. Arnold, 'Zur Frage der Milchversorgung der Städte', *Jahrbücher für Nationalökonomie und Statistik*, **96** (1911), 585–642.

P. Arnold/M. Sering (Hg.), *Milchwirtschaftliche Erzeugnisse*, München/Leipzig, 1912–1915 (Schriften des Vereins für Socialpolitik, Bd. 140, 1–5).

Beschreibung der milchwirtschaftlichen Verhältnisse im Deutschen Reiche, Berlin, 1895.

W. Beukemann, 'Der Milchverbrauch der Städte. Sein Zusammenhang mit den Landesverhältnissen und dem Wohlstande der Bewohner', in: *Bericht über die*

Allgemeine Ausstellung für hygienische Milchversorgung im Mai 1903 zu Hamburg, hg. v. d. Deutschen Milchwirtschaftlichen Verein, Hamburg, 1904, 95–130.
H. Brüning, *Geschichte der Methodik der künstlichen Säuglingsernährung. Nach medizin-, kultur- und kunstgeschichtlichen Studien*, Stuttgart, 1908.
A. Clevisch, *Die Versorgung der Städte mit Milch*, Hannover, 1909.
W. Engeling, *Handbuch für die praktischen Käserei*, Leipzig² 1901.
W. Fleischmann, *Das Molkereiwesen*, Braunschweig, 1875–1879.
W. Fleischmann, (Ed.): *Lehrbuch der Milchwirtschaft*, Leipzig, ⁴1908, ³1901, Bremen, ²1898, 1898.
E. Freund, *Milchwirtschaftliche Maschinen und Geräte*, Berlin, 1908.
Der Handel mit Milch und Molkereiprodukten. Verhandlungen und Berichte des Unterausschusses für Gewerbe: Industrie, Handel und Handwerk (III. Unterausschuß), 9. Arbeitsgruppe (Handel), Bd. 6, Berlin, 1929.
Heusner, 'Ueber Nutzen und Einrichtung der Milchcontrole in Städten', *Deutsche Vierteljahrsschrift für öffentliche Gesundheitspflege*, **9** (1877), 43–63.
O. Kamp, 'Der öffentliche Milchausschank in Rheinland und Westfalen', in: *Die gemeinnützige Milchversorgung in Deutschland*, München/Leipzig, 1914 (Schriften des Vereins für Socialpolitik, Bde. 140, 5), 111–164.
v. Klenze, *Handbuch der Käserei-Technik*, Bremen, 1884.
C. Knoch, *Die Magermilch-Verwertung in der Molkereien. Zusammenstellung der verschiedenen Verwertungsmethoden*, Leipzig, ²1912.
B. Martiny, *Die Milch, ihr Wesen und ihre Verwertung*, 2 Bde., Danzig, 1871.
Die Milch und ihre Bedeutung für Volkswirtschaft und Volksgesundheit. Dargestellt im Auftrage der wissenschaftlichen Abteilung der Allgemeinen Ausstellung für Hygienische Milchversorgung Hamburg 1903, Hamburg, 1903.
Die Milchwirtschaft in den verschiedenen Gergenden Deutschlands, hg. v. Milchwirtschaftlichen Vereine, Danzig, 1877.
v. Ohlen, 'Die Bekämpfung der Säuglingssterblichkeit durch öffentliche Organe und private Wohlthätigkeit mittels Beschaffung einwandfreier Kindermilch unter specieller Berücksichtigung Hamburger Verhältnisse', *Zeitschrift für Hygiene und Infektionskrankheiten*, **49** (1905), 199–281.
A. Schlossmann, 'Milchhandel und Milchregulative', in: P. Sommerfeld (Hg.): *Handbuch der Milchkunde*, Wiesbaden, 1909, 836–978.
R. Schreiber, *Die gesetzliche und polizeiliche Regelung des Milchverkehrs in Deutschland*, Rwiss. Diss. Erlangen, Kempten, 1912.
F. Soxhlet, 'Ein verbessertes Verfahren der Milch-Sterilisirung', *Münchener Medizinische Wochenschrift*, **38** (1891), 335–339, 353–356.
O. Trüdinger, 'Die Milchwirtschaft in Württemberg', *Württembergische Jahrbücher für Statistik und Landeskunde 1907* (1908), II.68-II.97.
J.F. Voigt, *Geschichtliches über die Versorgung Hamburgs mit Milch*, Hamburg, 1903.

Part III

Traditional Milk Products and Preparations

8 Cheese-making on the Åland Islands

Nils Storå

In the northern part of the Baltic Sea, the Åland Islands form a sort of island bridge between south-western Finland and Sweden (see Fig. 8.1). Although they belong to Finland, the islands and their 24,000 Swedish-speaking inhabitants enjoy autonomous status today. Their nineteenth century large fleet of collectively-owned sailing vessels has been replaced by modern equipment, cargo ships and tourist ferries sponsored by a few shipping companies concentrated in Mariehamn, the capital city. Workshop industries, and above all tourism, are important branches of the economy of Åland today, supplementing traditional means of livelihood such as farming, animal husbandry and fishing.

Tourists and holiday-makers are offered all kinds of products more or less typical of Åland, as well as traditional foods, such as Åland black bread (*Åländskt svartbröd*), Åland pancakes (*Ålandspannkaka*), and sometimes also Åland cheese (*Ålandsost*). In this context the phenomenon of revitalization plays an important role. Foods served as typical of Åland today have not, of course, remained unchanged over the centuries. Changes had taken place long before the revitalization process ever began. Nor are the foods necessarily typical of the Åland Islands alone. Western contacts particularly with Stockholm—the capital of Finland until its separation from Sweden in 1809—as well as eastern contacts with the Finnish mainland, have influenced the culture of the Åland islanders in many ways.

Today, the variations on the theme of Åland black bread all have a markedly *sweet* flavour, whereas eighteenth century sources describe this bread as being *sour*.[1] Today, also, the Åland pancake is served in a variety of ways, but none of them contains the ingredient which originally gave rise to its peculiar flavour, i.e. the large eggs of a particular species of sea-bird, the goosander (*Mergus merganser*), which gave this pancake a strong, rather fishy taste, and also its dark yellow colour. The type of cheese which is today served occasionally as Åland cheese is the only kind which is still homemade: it is the so-called 'sweet cheese' (*sötost*) a kind of soft, cake-like cheese, eaten fresh. Nowadays, sweet cheese is

98 *Milk and milk products*

usually prepared from full-cream milk, sometimes mixed with skimmed milk, by using sour milk as lactic acid. In earlier times the first milk taken from the cow after calving, the so-called 'raw milk', was commonly used, in which case no acid was needed. Since calving takes place in spring and early summer this was the main season for making sweet cheese. Although this cheese and a few other kinds of fresh cheese made on Åland,[2] do have

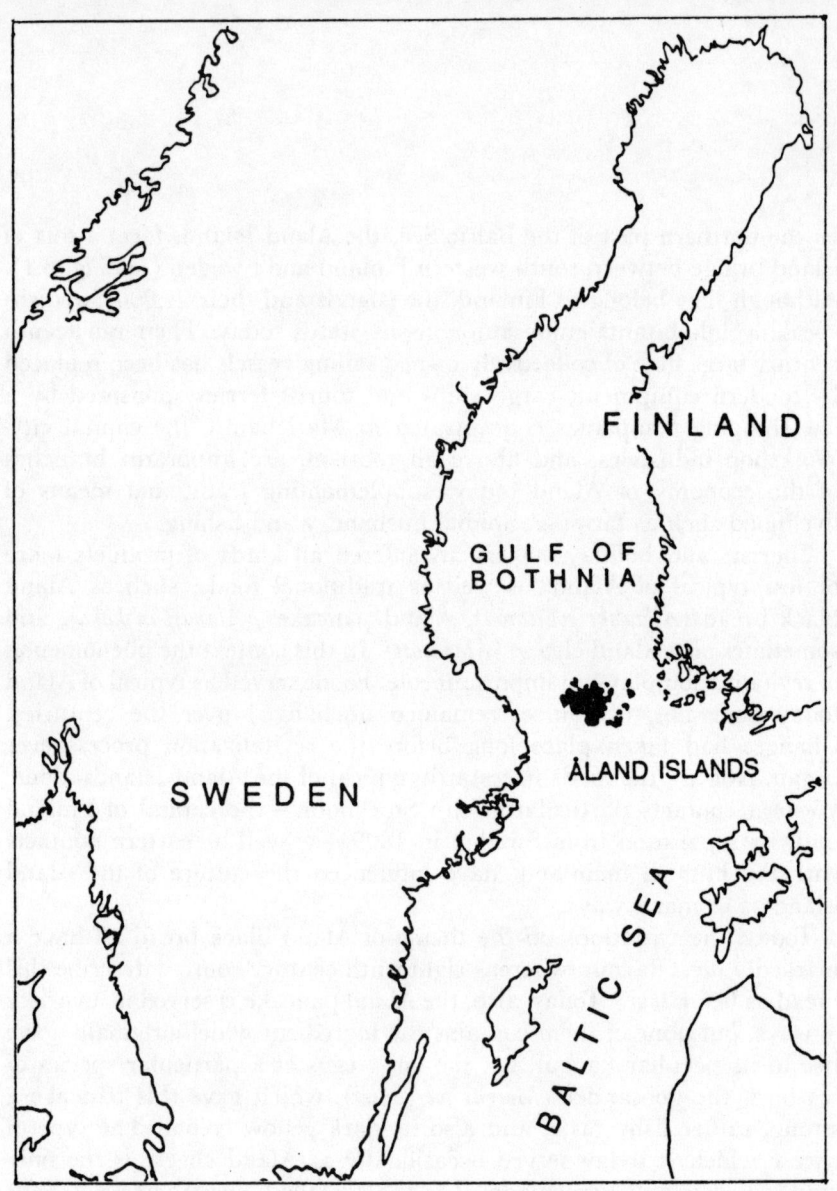

Fig. 8.1. The Åland Islands.

their place in traditional cheese-making, it was another type of cheese that made Åland peasant cheese known outside the islands.

Until the beginning of the twentieth century the concept of Åland cheese applied to a hard-pressed, tile-shaped cheese (Fig. 8.2), which was processed for storage. Unlike sweet cheese, which is boiled until very little whey is left, this was a type of true cheese, made from full-cream milk,

Fig. 8.2. Hard-pressed cheese from the Åland Islands.

with rennet used for coagulation, and ripened in storage. For about one kilogram (around two pounds) of cheese, ten to twelve litres (about two gallons) of full-cream milk—in more recent times mixed with skimmed milk—was required. The growing dairy industry gradually put an end to the production of this type of peasant cheese, which was known locally as mould-cheese (*stommanost*), from the rectangular, wooden moulds into which the curd was packed and from which the whey was discharged. After a day or two in the mould, the cheese was placed on a board or shelf and left to dry slowly in the open air for at least one month. Cheese for winter consumption was dried for up to three months. The final, slow ripening process took place in the granary where the cheese lay buried for months in the bins of rye before it was used. The cheese did not achieve its full flavour until it was a year old.[3] The long, tedious process of making this cheese contributed to its decline. The last cheese of this type was made in the 1920s. Apparently, the lengthy process has also been an obstacle—so far—to a modern revitalization of this sort of cheese.

In this context it is not necessary to go further into the process by which Åland mould cheese was made. I am more interested in finding answers to questions about why this kind of cheese was of such great importance to the islanders, and why Åland cheese enjoyed such a great reputation in Sweden and on the Finnish mainland. In this respect, mould cheese seems to reflect important elements of the Ålanders' culture.

The amount of full-cream milk required meant that cheese-making was especially concentrated in the households of farmers. Considerably less cheese was made by smallholders and fishermen. On the Åland Islands, winter feed for cattle was also scarce, which confined the lactation period proper, and thus cheese-making, to the summer months. This, of course, increased efforts aimed at producing cheese that could be stored for winter provisions. As has been rightly pointed out, cheese-making may be regarded as a method of preserving milk.

Of the early history of cheese-making on Åland very little is known, and the few early records hinting at Åland cheese offer no information about what *type* of cheese it was. We know that in the sixteenth century the milk used for cheese-making came mainly from goats and sheep, whereas the milk from cows seems to have been used primarily for the production of butter.[4] This early picture of the milk economy, however, applies to the three royal model farms in Åland, which were established by King Gustavus Vasa. There is reason to believe that these model farms played an important role in early Åland cheese-making.

According to Olaus Magnus (1555), who identifies certain districts in Sweden especially known for the production of cheese, goat cheese made in Finland had the most highly appreciated flavour of all. He also indicates that the cheese-makers in Finland knew how to enhance the flavour of smoked cheese by utilising the smoke of a strong-smelling plant (apparently *Ledum palustre*, 'skvattram', 'getpors'). Prepared in this way, the cheese could be stored for years.

Although Åland cheese is not specifically mentioned by Olaus Magnus it might well have been what he was describing. Until the nineteenth century, cheese-making in Finland was concentrated in the western part of the country, Åland included. There was some export of cheese from this area in the sixteenth and seventeenth centuries,[5] while the export of Åland cheese continued until the second half of the nineteenth century. Cheese was one of the many products of the Ålanders' trade, especially with Stockholm.[6] Judging by a number of early sources, Åland cheese was much favoured outside the islands, and not only in Stockholm, for its strong taste and richness of flavour. An attempt made in the 1750s to introduce the technique of Åland cheese-making into south-western Finland,[7] reflects the reputation of Åland housewives as skilled cheese-makers. From the point of view of cheese-making traditions, it is important to underline the fact that, due to invasions at the beginning of the eighteenth century, most of the Ålanders had to flee to Sweden where they stayed for eight years (1714–1722).[8] When they returned they brought new cattle with them, and possibly also new knowledge of how to make good cheese.

The first dairies on the Åland Islands were founded in the 1870s, and in the 1890s their number grew rapidly.[9] This naturally brought forth great changes in the peasant milk economy. The demand for home-made cheese diminished. At the beginning of the twentieth century dairies were established on a co-operative basis, which finally put an end to farm-based cheese-making. The rapid decline was publicly deplored at the agricultural exhibitions on Åland during this period. At the exhibition in 1887 few peasant cheeses were exhibited; in 1910 only six were shown.[10]

To what extent this short outline of the history of Åland cheese-making is also the history of mould cheese, is difficult to determine. The rectangular tripartite moulds used on Åland (Figs. 8.3–8.4) were not unique to this area. In his survey of traditional Swedish cheese-making, Gustav Ränk has pointed out that the same kind of moulds, consisting of 1) a 'cheese-board' (*ostränna*), (2) a loose frame or mould (*stomman*), and (3) a lid (*lock*), can be found in certain districts of Sweden, on the Åland Islands, and on some adjoining islands in south-western Finland, as well as among the Swedes in Estonia.[11] It appears from his detailed study that moulds of this kind were used as early as the Middle Ages. In his opinion cheese prepared with rennet has always been pressed in moulds.[12] Similar moulds, however, do not imply identical cheese-making processes.

The tile-shaped cheeses were, no doubt, easier to handle than round, cake-like, or square-shaped cheeses. During the process of drying them on boards and shelves, they had to be turned around every once in a while. In more recent times there has been no standard measure for

102 *Milk and milk products*

Fig. 8.3–4. Rectangular three-part Åland cheese moulds.

the moulds used: the size could differ to some extent from one farm to another. A more important advantage of using a mould was that the whey could be discharged effectively, which was done by pressing the lid by hand. The reason for the loose mould or frame was to enable the curd to be pressed from either side. The cheese-board had a hole in it through which the whey could flow into a vessel. Since every farm had its own marks or ornaments carved on the board and the lid, the origin of the cheeses could be identified. This was of special importance because for a long time cheeses were offered to the clergy[13] as a more or less obligatory gift. At certain festivals cheeses from different cheese-makers could be compared, and this led to competition among the peasant women, forcing them to look for special means of modifying the general pattern of the cheese-making process in order to get variations in the cheese.

Although the Åland Islands do not constitute a large area, there are many regional differences which are also reflected in cheese-making. In his description of Åland in 1795, F.W. Radloff states that the cheese of Åland which is known for its good flavour, actually comes from the island parish of Föglö, the only parish, according to him, engaged in the cheese trade.[14] Later, in 1852, some of the island communities in this parish are also identified as making the best Åland cheese.[15] This does not mean, however, that the Föglö cheese alone[16] would have given Åland cheese its good reputation. According to Radloff, cheese as good as that of Föglö could be found also in other places on Åland, but the only export from Åland stemmed from Föglö. The central position of this parish, on the sailing route between Finland and Sweden, apparently contributed to the local importance of the trade in cheese and many other products. Besides, animal husbandry was for a long time more important in this parish than in many others on Åland.[17]

Although cheese was exported from the Åland Islands, production was mainly geared towards local consumption as it was one of the most important foods. Mould cheese was generally put on black bread instead of butter, particularly during haymaking, harvesting and other periods of hard work, as well as on some festive occasions. Whey, mixed with sour milk, was the most common food during haymaking, the first period of hard work following the traditional time of cheese-making. Above all, however, this cheese, combined with black bread, was consumed during summer and winter travel. It was, therefore, part of the standard meals provided by the farmers for the crews of their collectively-owned sailing ships.[18] This cheese had the great advantage of not freezing in winter. Cheese and black bread thus constituted the necessary provisions for seal-hunters and others making long journeys on the ice. From the beginning of the seventeenth century, a large portion of the male population on Åland was engaged in the transport of mail between Finland and Sweden, which, until the 1860s, involved travelling both on ice and on the open water.[19] When collective seafaring came to an end towards the end of the nineteenth century, the need for cheese for traditional purposes gradually diminished—at the same time as the local dairy industry increasingly replaced home cheese-making.

On the Åland Islands mould cheeses were stored in the bins of rye for several months. The ripening process went on during storage, and the final product was a hard, dry cheese, thick enough, however, to preserve some softness in the middle. In Sweden, the cheese pressed in moulds was later dried in much the same way as Åland cheese, and in some districts in Sweden, too, the cheese could be stored in bins of grain,[20] but here there was a significant difference as compared to the Åland storage in rye-bins. As on the mainland of Finland, the rye in Åland had been dried in special drying kilns which gave the grain the flavour of smoke—and the cheese could not be buried for months in the smokey rye without having its flavour affected. Therefore, there is reason to believe that this particular storage-place contributed to the flavour of Åland cheese. There is no evidence in Åland for the practice of directly *smoking* cheese, as related by Olaus Magnus.

As a whole, the Åland Islands constitute an area with unusually rich flora.[21] Although there are regional differences, its good soil and mild climate have brought forth vegetation well suited for the grazing of cattle. This seems to be one of the reasons for the high reputation of Åland cheese. The rich herbs of the grazing grounds contributed to the flavour of the cheese, as the occasional early visitor and writer has also pointed out.[22] There is no evidence for the use of herbal rennet on the Åland Islands, although a number of herbs used as such elsewhere[23] do belong to the flora of Åland.

Åland mould cheese was prepared with very little, if any, salt,[24] apart from the salt from the traditional rennet made from the mawskin of a young calf and always salted. The *curd* seems never to have been salted. The cheese could sometimes be rubbed over with salt, or it could be put into brine for some time. If the surface grew mouldy, it was rubbed with juniper or washed with juniper water, which also affected the flavour. It may be surmised that the little salt used partly reflects the scarce supply of drinking water on board the sailing vessels, where thirst was a constant problem.

As we have seen, the particular flavour that made Åland cheese much favoured, may be explained by several factors. Its strong taste compared to other cheese could have been the result of mixing some goat's milk with the full-cream milk of cows. There is evidence for this practice still in the 1890s,[25] although the proportions of the two kinds of milk are not known. It is quite obvious that the last peasant cheeses were prepared in districts where goats were still kept (to some extent). In the far-off island community of Kökar, true goat cheeses have been made within living memory.[26] In the central agricultural districts of Åland the goats disappeared as early as the eighteenth century, mainly because of the difficulties in keeping them away from the fields, whereas goats continued to be used in the outer, barren islands, where this was no real problem. Goat milk was added in order to make the cheese 'stronger and fatter', as witnessed also in areas outside Åland.[27]

Why cheese-making has had a long tradition only in western (and northern) Finland, and not in the eastern part, where sour milk of several kinds, but no real cheese, was traditionally made, has been explained as diffusion from Sweden, where the closest parallels have been found. Considered from a diffusionist perspective, it is easy to regard the Ålanders as the transmitters of cultural elements and ideas from Sweden to Finland. This may have been the case with mould cheese, but as I have attempted to demonstrate, with the exception of the storage in kiln-dried grain, which probably was an important part of the cheese-making process, the fact that the Ålanders dried their grain in smoke-kilns, according to an eastern Finnish practice, made the conditions of storage different from those in Sweden. Thus, mould cheese, like other cultural traits, shows that the Ålanders could take advantage of their location, combining and successfully applying ideas from the west and the east to their local resources. The mobility of the Åland population, resulting from sea-faring and other circumstances, made this possible.

This paper is partly based on the results of a questionnaire (No. 18) sent out by the Department of Ethnology, Åbo Akademi University, in 1966. Among the informants contributing were 28 from the Åland Islands.

NOTES AND REFERENCES

1 F. W. Radloff, *Beskrifning öfver Åland*, Åbo 1795, 245.
2 A-L. Dreijer, 'Osten i den åländska allmogehushållningen', *Åländsk odling*, 1967, 106ff.
3 G. Grotenfelt, *Vanhanaikainen suomalainen maitotalous*, Helsinki, 1916, 152.
4 K. A. Bomansson, H. A. Reinholm, *Finlands Fornborgar 1: Kastelholm*, Helsingfors, 1856, 149.
5 U.T. Sirelius, *Suomen kansanomaista kulttuuria* 1, Helsinki, 1919, 333.
6 N. Friberg, *Stockholm i bottniska farvatten*, Stockholms universitet, Kulturgeografiska institutionen, Meddelanden B 59, Stockholm 1983, 68f; C. Ramsdahl, *Det åländska folkets historia* II: 1, Mariehamn, 1988, 182.
7 G. Grotenfelt, *Bidrag till kännedomen af mjölkhushållningens utveckling i Finland*, Kuopio, 1906, 90f.; *op. cit.*, 1916, 150ff.
8 N. Storå, 'Åland—kulturområde mellan väst och öst', *Ålands högskolas skriftserie* I, Mariehamn, 1988, 19.
9 U. Forss, *Andelsmejerierna i Finlands svenskbygd*, Borgå 1958, 17.
10 N. Storå, 'Folkkultur i Föglö', *Åländsk odling*, 1989, (1990), 30.
11 G. Ränk, *Från mjölk till ost*, Nordiska museets handlingar 66, Lund 1966, 124; cf. R. Ahlbäck, *Kulturgeografiska kartor över Svenskfinland*, Svenska litteratursällskapet i Finland, Folklivsstudier 1, Helsingfors, 1945, 122ff.
12 G. Ränk, *ibid.*, 114.

13 Storå, *op. cit.*, 1990, 30.
14 F. W. Radloff, *op. cit.*, 1795, 199.
15 M. Weckström, *Geografiskt-statistiskt lexikon öfver Finland: Åland*, Åbo, 1852, 21, 32.
16 cf. S. Kärkkäinen, 'Föglöosten', *Åländsk odling* 1989, Mariehamn 1990.
17 Storå, *op. cit.*, 1990, 25f., 32.
18 N. Storå, 'Adaptive Dynamics and Island Life', *Studia Fennica* **30**, 1985, 123f.
19 N. Storå, 'Havet som skiljer och förenar', *Glimtar ur Ålands folkkultur* **3**, Mariehamn 1988, 7ff.
20 Ränk, *op. cit.*, 1966, 138.
21 W. R. Mead & S.H. Jaatinen, *The Åland Islands*, Plymouth 1975, 48ff.
22 Storå, *op. cit.*, 1990, 30.
23 A. Olsson, 'Koagulerad mjölk i äldre tiders hushållning', *Rig* **44**, 1961, 117ff.
24 W. Solstrand, 'Etnografiska anteckningar, Brändö (Åland)', *Hembygden*, 1913, 8.
25 K.E.F. Ignatius, *Finlands geografi*, Helsingfors 1891, 534; Olsson, *op. cit.*, *1895, 41.*
26 R. Ahlbäck, *Kökar. Näringslivet och dess organisation i en utskärssocken*, Helsingfors, 1955, 269.
27 G. Grotenfelt, *op. cit.*, 1906, 93; G. Grotenfelt, *Suomen vuohikirja*, Porvoo 1992, 156ff.

SELECT BIBLIOGRAPHY

A. L. Dreijer, 'Osten i den åländska allmogehushållningen', *Åländsk odling 1967*, 102-120.

R. Forsen, *Die Langmilch*, Helsinki, 1966.

G. Grotenfelt, *Bidrag till kännedomen af Mjölkhushållningens utveckling i Finland*, Kuopio, 1906.

G. Grotenfelt, *Vanhanaikainen suomalainen maitotalous*, Helsinki, 1916.

S.-B. Kärkkäinen, 'Föglöosten', *Aländsk odling* 1989 (1990), 45–68.

J. Talvi, *Rakastettu juusto*, Keuruu, 1981.

T. Vahter, 'Varsinais-Suomen juustoa ja juustokehiä, *Kotiseutu*, 1935, 20–27.

K.Vilkuna, 'Makeajuusto', *Kotiseutu* 1938, 129–134.

K.Vilkuna, 'Der am Feuer gegrillte Käse', *Schweizerisches Archiv für Volkskunde*, 1972–1973, 694–702.

H. Vilppula, 'Finsk ost', *Fataburen*, 1946, 83–92.

9 Curds and pressed cottage cheese in Latvia

Linda Dumpe

The unceasing and variable interaction of different ethnic communities has left noteworthy impressions on traditional European culture. They can be best observed in the regions with intensive ethnic contacts, and in those domains of culture that maintain stable traditions. The eastern Baltic area deserves special attention because of the lasting interaction of various cultures. The subject of my paper is the manifestation of this interaction in traditional dairy farming, a branch of popular culture with distinctly stable traditions. The character of dairy farming in the pre-industrial age depended on the general purpose of a farm, which in its turn was determined by ecological conditions.

The moist and temperate maritime climate and rich grass ensured favourable conditions for cattle breeding in Latvia. Thus dairy produce has always been an essential part of Latvian peasants' food. At the same time, certain trends in agriculture limited the development of dairy farming. As the main branch, tillage farming, required obligatory manuring of fields, the most important task was the collecting of manure; dairy produce was of secondary significance, being only for consumption, and that determined the system of milk processing. The milk yield of the extensive dairy herds was low, so milk processing was based on the gradual collection and natural souring of milk. The processing of milk based on souring was the natural one in the temperate climate of Latvia; milk turned sour slowly, and isolation of fats and protein parts was satisfactory.

The basic phases of milk processing in Latvia were: (1) the settling and souring of milk; (2) the skimming of cream from the milk; (3) collecting the sour cream and churning (butter was regarded as the most valuable dairy product, which was stored, and considered a commodity since the Middle Ages, (4) the processing of skimmed sour milk into curds; (5) the processing of curds into cheese. During the periods of extensive agriculture, a similar system of milk processing and dairy farming had embraced the whole East European Plain, from the Carpathians in the south to Scandinavia in the north. The similarity of milk processing

108   Milk and milk products

techniques in such a vast area populated by ethnically diverse peoples, can be explained historically in terms of dissemination from the south together with cattle breeding.[1]

The spread of dairy farming to the north brought certain alterations to the processing of milk due to local ecological conditions, particularly in the Baltics. This paper is devoted to the final stages of dairy processing, i.e. the making of curds and pressed cottage cheese, in Latvia, because in this process the influence of specific local traditions can be seen.

A short introduction is necessary in order to understand the role of curds and cottage cheese in the general context of European dairy culture. These products are among the most ancient dairy products in Europe. In the Baltic region this is manifested in the cognate designations for cheese: Latvian *siers*, Lithuanian *suris*, Old Prussian *suris*, Russian and other East Slavonic languages *cbip*, Estonian *sõir*, Livonian *seir*, Votian *sura*. These etymologies can be complemented by Old Norse *súrr* (sour), *sýra* (sour milk), and Old High German *sur* (sour).

Curds are a popular dairy product, and it is still consumed in different ways in Europe. The production of cottage cheese, on the contrary, has undergone fundamental changes. In the 1st millenium AD, fermented milk cheese made its way into Europe from the nomadic peoples of the Near East through Greece and Rome.[2] The fermented cheese had better taste and preservation qualities, and it thus spread quickly and forced out cottage cheese. It appeared in the Baltic towns, abbeys and estates not later than in the sixteenth century.[3] Already in 1584, the Bishop of Cesis, Otto Schenking, invited a Dutchman to Latvia to make Dutch cheese. Later the Dutchman stayed in the service of the local Jesuits. However, the fermented cheese did not take root among Latvian peasants, and until the 1950s they produced and consumed essentially only cottage cheese.[4]

There are two basic methods of making curds, and these methods have specific regional distributions: in the western part of Latvia (Kurzeme) curds were made by intensively souring milk until the whey separated. The resulting protein substance was an acid product of soft consistency, something between sour milk and curds. The method of making curds by means of intensive souring is considered the primary one in north-eastern Europe.[5] In the nineteenth century it was, however, replaced by a more effective method—the warming of sour milk. Thus the eastern coasts of the Baltic Sea where intensive souring of milk is still remembered can be regarded as the relic areas of this method.

In eastern Latvia (Latgale), according to ethnographic materials, curds were made only by the second method, i.e. warming. Evidently this method of curd production is a secondary one in the northern part of Eastern Europe, since it spread with the Slave migration northwards, together with their peculiar hearth, the cooking stove.[6] In addition to the stove, methods of food preparation in stoves were also borrowed. Thus curds were also made by putting a vessel of sour milk into a warm stove. The method of warming sour milk to make curds took root in the eastern Baltic area, and this cultural loan from the Slave territory is proved by linguistic material: the designation of curds in eastern Latvia is *tārags, tvorogs*—a borrowing from Russian творг. A similar origin can be traced in Lithuanian—*varške*, and in eastern Estonia—*varok, vorok*.[7] Curds made by warming required less sour milk, so the product was not acidic. Moreover, it was harder, and could be preserved for a longer period, and that is why the warming method spread beyond the distribution area of the cooking stove. In the second half of the nineteenth century, making curds by warming sour milk became known in the western part of Latvia. The cottage cheese made in Latvia was of two kinds: the warmed soft *Jāņu siers* (John's Day cheese) and the hard (pressed) one.

The basic method of making soft cheese was to warm curds in milk. The milk also turned sour, and the jelly-like substance was isolated, and eggs, butter, cream and caraway seeds were added. It was warmed and kneaded until it acquired a homogeneous consistency. The designation of this particular kind of cheese-making (*siera siešana*) was derived from the process itself: the substance was put into a cloth, the corners of the cloth were fastened above the 'slab' of cheese, and it was slightly pressed. In that way the slab could acquire either round or square form. The cheese was ready for consumption in a few hours, but it could not be preserved for long. This type of cheese was made in the western and central parts of Latvia.

In eastern Latvia, another type of cheese—hard-pressed cheese—was common. Curds were kneaded with salt, caraway seeds, and sometimes also with cream and eggs. The substance was put into a linen cloth, and then into a special wooden appliance for pressing. The ripe cheese had a flat round or elongated form. If it was intended for longer storage it was dried in the wind (usually in a punched box, basket or cloth). Sometimes cheese was smoked or lightly baked.

As we can see, the areas of both cheese types roughly correspond to the respective areas of curd types. Thus we can establish a connection between cheese types and the respective modes of curd-making.

Curds made by intensive souring were soft and had an acidic taste. In cheese making, they could be used only for curdling fresh milk: it was in this way that the soft warmed cheese was made. The warmed curds of more solid consistency could be used for making pressed cheese.

We must take into account the spread of both Latvian cottage-cheese types in north-eastern Europe. Similar soft warmed cheese was also made in southern Estonia,[8] and with slight differences in the southern and central districts of Finland, Sweden and Norway[9]—in general terms it spread in a rather limited area around the Baltic Sea. The solid pressed cottage cheese characteristic of eastern Latvia was common in a vast East European area to the south and east of the Baltics.[10] Thus, the comparative material shows that the different methods of making curds and cottage cheese in Latvia, belonged to two regions; one of them was located in the north-western part of Eastern Europe, the other in the south-eastern part, with the Baltics as the contact zone for both regions. In Latvia, the borderline between the regions separates its western and central parts from the eastern one. The elements of the region's ancient dairy farming are still preserved in western Latvia. The relationship between the milk processing techniques in eastern Latvia, and the south-eastern region of dairy farming, can be explained by Slavic influence.

Several other types of curd cheese were made in the Baltic Sea region. In Latvia, the northern German *knappkäse* and beer cheese were quite widespread; they were introduced by German immigrants living in estates and towns in the Middle Ages, and are mentioned also in documents of that time.[11] In the eighteenth and nineteenth centuries, curd cheese of the northern German type were relatively often used in the Baltic German estates[12] and to a much lesser extent among peasants also. The name of the cheese was derived from the German language—*knapsieriņi* (Latvian)—*Knappkäse* (German). Thus in characterising the types of curd cheese in Latvia, a certain Central European curd cheese influence from the Middle Ages can be seen.

Finally, a few words about cheese in Latvian customs, because the antiquity of the cheese culture in the Baltics is proved by its extensive function in rituals. As church visitation records show, cheese was included in sacrificial rites even in the seventeenth and eighteenth centuries.[13] According to folk beliefs, cheese was made to feed the souls of deceased ancestors, and it was also an integral element of all important moments of human life, and of popular festivals. But the major role that cheese played was at the summer solstice—St John's Day. The ancient connection between warmed soft cheese and the summer solstice rites is also manifested in its most widespread name in Latvian: *Jāņu siers* (John's cheese). The round or multi-angular shape and yellow colour of cheese unmistakably presents evidence of a solar cult. Cheese is one of the chief elements in the St John's Day festival. It is known that the solstice customs were looked upon as possessing special power. A motive for the cheese tradition on St John's Day in Latvia is closely connected with producing magic rites associated with cattle breeding. According to earlier ways of thinking, peasants considered that ritual beneficial for their cattle and milk yield.

The following is a short survey of the main customs connected with the Latvian summer solstice and their significance.

1. Making cheese was the responsibility of the farmer's wife. She also offered cheese to the guests. Latvian folk songs contain many lines dedicated to the farmer's wife, or *Jāņu māte* (John's mother), the name closely associated with the summer solstice rite. She is customarily said to be meeting participants of the St John's Day festival, revellers of *Līgo* night, with round cheese in her apron in a basket. The direct link between the ritual and farmer's wife is obvious, because of her duty to take care of dairy cattle.

2. Cheese had to be available in abundant quantities. That was also mentioned in folkore, in sayings like 'to carry cheese in a sheet', 'to build a stove of cheese'. Consequently, the round piece of cheese had to be big as well—sometimes it is compared to a mill-wheel. As elements of imitative magic 'many cheeses', and 'big cheeses', could further the yield of dairy produce.
3. St John's Day cheese was round or multi-angular, triangular, hexagonal, etc.. Usually it was yellow—'as round wax'. In folklore, one also finds reference to white cheese connected with the wish to have calves with white heads, while multi-coloured cheese was associated with the desire to have 'only dappled calves'. Here again elements of imitative magic are obvious.
4. As cheese is mainly associated with the idea of cattle well-being, it was customary that on the solstice evening (in Latvia: *Līgo vakars*), cheese was offered to shepherds with accompanying sayings such as: 'many cows yield a lot of milk'. There are many folksongs in which shepherds ask for cheese, and they threaten not to tend the cows, or to leave them in the cattle-shed, if their request is not granted.
5. There existed a custom to meet neighbours at the time of the summer solstice, and to walk around in a crowd at night singing special *Līgo* songs. When guests arrived at a farmer's house they were treated to cheese—as a gesture of gratitude for singing and bringing wild flowers. Stingy housewives, on the contrary, received threats that their cows would be bewitched, or that a wolf would be let into the cattle-yard.
6. The solstice night was celebrated in Latvia by making fires on many hills and hillocks, or burning barrels of tar on poles—in general, making *Jāņu ugunis* (John's fires). They were intended to scare away various evils. People also ate cheese by the fires.

Observing cottage-cheese rituals in Latvia and in the surrounding territory of north-eastern Europe, it should be noted that making cottage cheese for the summer solstice celebration is characteristic of other countries with developed dairy-farming in the Baltic region, such as Lithuania, southern Estonia, Finland and Swedish districts with developed dairy-farming.

Motivations for such cheese traditions in those territories, including Latvia, are the well-being of cattle and an increase in the milk-yield; however, the scope and stability of Latvian cheese-rituals are superior to those found elsewhere.

In conclusion, while Latvia belongs to the common region of cottage cheese in north-eastern Europe, nevertheless cottage cheese-culture in Latvia has the following special features: (a) Latvia is a contact area of two types of cottage cheese; (b) the soft warmed cottage cheese is made in a special way in Latvia, and (c) a stable ritual involving the soft warmed cottage cheese has been maintained till the present time.

112   Milk and milk products

These features allow us to regard Latvia as a variable area of the north-eastern Europe cottage cheese region, and it must be taken into account when considering the historical and geographical aspects of European dairy-farming culture.

NOTES

1. G. Ränk, 'Verbreitungsverhältnisse einiger Milchprodukte im Eurasischen Raum', *Ethnologia Scandinavica*, 1971, 125.
2. F. Anderegg, *Allgemeine Geschichte der Michwirtschaft*, Zürich 1894, 12–17. G. Ränk, 'Gegorene Milch und Käse bei den Hirtenvölkern Asiens', *Suomalis-ugrilaisen sueran aikakauskirjasta*, **70**, 1969, 56–66.
3. H.v. Bruiningk, *Annalekten zur Geschichte der Milkwirtschaft und der Viehzucht in Livland*, Riga, 1907, 9.
4. The adaptation process of fermented cheese in Latvia is revealed by data acquired by the author from an inquiry into the consumption of dairy products in the countryside of Latvia in 1989. Of three hundred and sixty families surveyed, forty-four did not buy fermented cheese, or bought it very rarely; they used mainly curd cheese. The most interesting aspect is the attitude of different age-groups towards fermented cheese. Ninety per cent of those over sixty mentioned that they prefered curd cheese; people in the forty to fifty age-group used both types of cheese; only people under forty years of age preferred fermented cheese to curd cheese.
5. G. Ränk, 'Om äldre mjölkhushållning i Balticum', *Svio-Estonica* **13**, 1956, 171–172; A. Moora, 'Über die volkstümliche Nahrung in der ostestnischen Grenzzone', *Congressus secundus internationalis fenno-ugristarum 2*, Helsinki 1968, 230.
6. K. Moszyński, *Kultura ludowa Słowian. I Kultura materialna*, Warszawa 1967, 285.
7. A. Moora, *op. cit.*, 230; G. Ränk, *op. cit.*, 1956, 172.
8. G. Ränk, *op. cit.*, 1956, 186.
9. G. Ränk, *Från mjolk till ost*, Stockholm 1966, 53–57.
10. L. Niederle, *Slavjanskije drevnosti*, Moskow 1956, 198; R. Merkienė, *Gyvulių ūkis XVIa-XXa pirmajoje pusėje*, Vilnius 1989, 111–116.
11. A. Soom, *Der Herrenhofin Estlandim /7. Jahrhundert*, Lund 1954, 127.
12. W. Hahn, *Die bäuerlichen Verhältnisse auf den herzöglichen Domänen Kurlandsim XVII und XVIII Jahrhundert*, Karlsruhe 1911, 22; J. Chr. Brotze, *Sammlung verschiedener Liefländischen Monumente Prospecte, Münzen, Wappen etc.*, Bd. 8, 88. (Manuscript from the collections of the Fundamental Library of the Latvian Academy of Sciences).
13. K. Bregžis, *Baznīcu vizitāciju protokoli*, Rīgā, 1931.

SELECT BIBLIOGRAPHY

W. Adolphi, *Pamācīšana moderēm, kā pie lopu kopšanas būs turēties*, Jelgava, 1837.
M. Amols, *Piens, sviests un siers un viņu apstrādāšana*, Jelgava, 1899.
A. Bielenstein, *Die Holzbauten und Holzgeräte der Letten*, Bd. 2, Petrograd, 1918.
O. Dārznieks, *Piena izmantošana*, Rīga, 1928.
L. Dumpe, *Lopkopība Latvijā 19. gs.-20.gs. sākumā*, Rīga, 1985, 57–87.
L. Dumpe, 'Etnokulturnije svjazi latišei i sosedņih narodov po dannim ekstevsivnogo moločnogo životnovodstva', *Problemi etnogenezai etņičeskoi istorii baltov*, Vilnius, 1985, 248–258
L. Dumpe, 'Baltisch-slawisch-deutsche Kulturströmungen. Dargestellt am Beispiel

der Milchwirtschaft im Ostbaltikum', *Jahrbüch für Vokskunde und Kulturgeschichte*, Bd. 32, 1989, 82-88.

L. Dumpe, 'Etničeskaja tradicija v moločnoi pišče seļskogo naseleņija Latvijskoi SSR', *Etnokuļturnije tradiciji sovremennostj*, 1989, 109-116.

K. Elberfeld, *No govju lopiem, kātos būsaudzināt, kopt un slimības dziedināt*, Jelgava, 1810.

J. Feldt, *Die Thierproduktion in Kurland*, Riga, 1865.

C. Jordan, *Praktisches Handbuch der Rindviehzucht, oder vollst ändige Anleituug zur Zucht, Plege und Nutzung des Rindes*, Dorpat, 1852.

P. Kārkliņš, *Eksportsviesta pagatavošana*, Rīga, 1929.

E. Kliesmets, *Meņšikovs N. Sieru ražčšana*, Rīga, 1965.

Lavijas piensaimnieku Centrālā Savienība. Gada darības pārskats, Riga, 1925-1939.

Latviska pavāru-grāmata. Muižas pavāriem par mācību, visādus kungu ēdienus gardi vārit un sataisīt, Jelgava, 1796.

K. Lielgalvis, *Siernieciba*, Rīga, 1930.

G. Mangaliņš, *Pratiskā piensaimniecība*, Rīga, 1928.

N. G. Meņšikov, 'Očerk razvitija maslodelija i sirovarneija v Latvii', *Sbornik trudov Latvijskogo naučno-issledovateļkogo instituta životnovodstva i veterinarii*, IX, 1958, 251-274.

Fr. Neilands, *Piens un piena rūpniecība*, Rīga, 1946.

Piensaimnieku sabiedrības 30 gadu darības atcerei, Rīga, 1939.

G. Ränk, 'Om äldre mjölkhushållning i Baltikum', *Svio-Estonica*, 1956, vol. 13, 165-200.

M. Redelien, *Haus und Herd*, Riga, 1895.

P. Rubenis, *Krējuma siers. Praktiska rokas grāmatiņa krējuma jeb baksteina siera izgatavošanai*, Jelgava, 1908.

P. Rubenis, *Mazais sviestnieks*, Peterburga, 1911.

P. Rubenis, *Mazais sviestnieks, Praktiski aizrādījumi sieru pagatavošanā*, Rīga, 1927.

A. Skroders, *Mīkstie sieri*, Rīga, 1926.

P. Staģis, *Latvijas lopkopības pārraudzības biedrību panākumi 30 gagados (1904-1934)*, Rīga, 1935.

J. Stankevičs, 'Latvijā ražoto sieru ķīmiskais sastāvs, Latvijas Universitates Raksti, III, 3, 1936, 105-240.

M. Strade, *Lopkopība un piensaimniecība*, Rīga, 1908.

A. Valdmanis, *Piensaimniecība*, Rīga, 1929.

E. Zariņš, *Piens un pien a preparaāi*, Rīga, 1922.

10 Variations on gomme

Karin Kvideland

The subject of this paper is one of the many old milk products of Norway. It has been known in Norway for several centuries and has various names but it is generally referred to as *gomme*. The nearest English equivalent is probably 'curds'.[1]

Three types of this product are treated here—*gomme* from the island Finnøy in the province of Rogaland, Southwest Norway, *gomme* from Upper Setesdal, in the province of Aust-Agder, South Norway, and *gomme* from the province of Troms, Northern Norway. My sources are as follows: the traditions about *gomme* which I learned in the home of my mother-in-law, in Reilstad on Finnøy, where it has been a traditional food for several generations; Gro Nomeland Gjerden's[2] work on old food traditions in Upper Setesdal; and Kristine Lind Grøttland's[3] work on the foodways of Northern Norway. Both Gjerden and Grøttland were born in the region they describe and where they collected the traditions, and they cover the period from the second half of the last century to our own time. In spite of similarities, Setesdal and Northern Norway differ with regard to food production, since the former is an inland valley, while the latter is a coastal area and dependent on seasonal fishing in the archipelago of Lofoten in the province of Troms, and on the fishing grounds of Finmark. On the island of Finnøy people live by farming and fishing.

WHAT IS GOMME?

Gomme has been variously regarded as a milk or a cheese product, and this ambiguity is not surprising in view of the method of its preparation and its consistency, which are detailed below. As in the production of actual cheese, the basic ingredient is milk, and the method of its preparation follows the cheese-making process up to a certain point—until the milk curdles and whey and curds form. It is then further treated in a variety of ways which differ from those used in cheese-making, in order to finish the product. Then, depending on whether it is for immediate household use, or perhaps for festive or commercial use, it might be decorated, or preserved and made ready for transportation.

As we shall now see, one of the product's characteristics is variation; the name, ingredients, ways of preparation, the serving, and the occasions on which it is served, all vary between the regions and within the regions.

TERMINOLOGY

The two most common names for this old Norwegian milk product are *gomme* and *dravle*, terms found in old Norwegian dictionaries whose compilers pursued topographical interests, and who also provide explanations of Norwegian dialect words characteristic of the area where they worked. The first printed Norwegian dictionary, *Den Norske Dictionarium eller Glosebog*, compiled by the minister Christen Jensön, (Copenhagen, 1646)[4], explains *gombe* (referring to Sundfjordmål) as a kind of cheese made from fresh milk. The word *dravle* does not occur in Jensön's dictionary.

In the 1740s Knud Leem,[5] a minister at Avaldsnes church on the island Karmøy (Vestlandet) explains *gombe* as a 'kind of dish made from fresh milk which is cooked until it has become a thick porridge, grøt, or cheese, used by the peasants at their weddings instead of cakes' (*bakkelse*.) As a second name for the dish, Leem gives the term *søost*, used in Sundfjord, and the explanation is the same. In 1749 the *Glossarium Norvagicum*, of Erik Pontoppidan, bishop of Bergen, was printed.[6] This booklet has the term *draule* which is glossed as follows: 'milk which is cooked as thick as butter or cheese. It was a great peasant delicacy'. *Gomme* is not found in Pontoppidan's glossary.

In 1850 Ivar Aasen's *Norsk Ordbok*[7] appeared. About *dravle* he says: 'a milk dish. Milk is cooked so that the cheese is separated from the whey and coagulates in hard lumps. Mostly used in the Western and Northern parts of the country. In Østerdalen *dravle* is used for a better quality cheese from fresh milk'. Aasen also has the term *gumbe* meaning a red-cooked cheese from fresh milk, i.e., sweet cheese (and he gives additional explanations of related terms).

Norsk ordbok[8] (1978) gives the following four meanings for the term *dravle*:

1. A dish made of curdled milk, cooked until the cheese has formed lumps, often eaten with sour cream. Provenance: Hallingdal, Lista, Rogaland, Björgvin bispedømme, all west Norway and Nordenfjells which refers to the areas north of the Dovre mountains in Trøndelang, Northern Norway.
2. Curds, and for this four meanings are given:
    a) A product from cheese-making, i.e. heating milk so that it curdles.

116 *Milk and milk products*

    b) Small lumps of curd remaining in the pot after cheese-making.
    c) Small-grained foam floating up during cheese-making.
    d) Curds floating up during the cooking of whey after the cheese has been taken out.
3. Cheese, or a dish similar to cheese, made by cooking milk, buttermilk and whey:
    a) A kind of fine cheese from fresh milk.
    b) Cheese from first milk (*råmjølk*); boiled-down curdled first milk.
    c) Boiled-down sweet whey.
    d) Cheese from boiled-down sour milk; soft cheese from boiled-down buttermilk (or milk) and whey; cheese from buttermilk; *red-boiled haglette*.
4. Gruel from buttermilk cheese with flour.

Nynorskordboka[9] (1986) has two meanings for *dravle*:

    a) A milk dish
    b) A name for different types of cheese made of boiled-down milk, buttermilk, or whey.

The same dictionary explains *gumme* (a word of uncertain origin) as:

    a) A kind of cheese from curdled milk.
    b) *Dravle*

Thus the terms *gomme* and *dravle* are both used for a milk dish as well as for a cheese product; in fact both terms refer to such a variety of often overlapping milk-based preparations that a certain ambiguity in the definition of the product can hardly be avoided.

    Other names for the milk product such as *gumme-ost*, and *gumbe-graut*, are compounds arising from the use of the terms *gomme* or *dravle* with the Norwegian words for cheese, (*ost*) and porridge (*grøt*), together with the adjectives 'white' as in *kvitgumme*, and 'red' as in *raudravle*, indicating the colour of the final product—which can vary between light yellow, through golden to dark brown.

    There are also other compound names, such as Nordfjord-*gomme*, *Jærgomme*, resulting from the combination of *gomme* with a geographical designation.[10] The dish is also known under names unrelated to either *gomme* or *dravle* such as *haglette*, *kabrette* and *søost*.

    In 1925 Nils Lid, Norway's first professor of ethnology, published a note on *gomme*. Lid's information stems from informants born in the second half of the nineteenth century, and among the provinces he mentions is Rogaland, where the island of Finnøy—of relevance for this paper—is also situated. Nils Lid's informant from Rogaland also reveals the varying names. The informant says: '*Gomme* is made by boiling milk. At boiling point buttermilk is added, and the *dravle* i.e. the curds, are boiled together with the whey until the *dravle* has turned a reddish brown'.[11] For present purposes, therefore, it suffices to sum up the situation in relation to terminology by stating that there are, apart from regional dialect terms, two main names for the milk product under discussion here—*dravle*, going back to Old Norwegian, and *gomme*, a word of uncertain origin.

INGREDIENTS AND PREPARATION

The Finnøy-*gomme* consists of sugar for caramel, milk (today pasteurized), sour milk, eggs and sugar for separating the curds and whey, as well as for thickening. My mother-in-law taught me to start with 500 grams sugar and to caramelize it. Into the pan with the caramelized sugar I should then pour 10 litres of milk and bring it to the boil. Two litres of sour milk are mixed with 10 eggs and 500 grams of sugar. This mixture is added to the hot milk. The milk curdles, and the lumps and the whey are boiled down to the desired brownish colour, which takes from about four to five hours; the resulting *gomme* consists of big porous lumps. The thickening agent to be used is made from 3 eggs, 3 tablespoons of sugar, and 4 tablespoons of flour. These ingredients are mixed in a separate pan, whey is added, and all is cooked for five to eight minutes 'to take away the taste of flour', as my mother-in-law used to explain. This thickening mixture is then stirred into the *gomme*; the dish is removed from the heat and left to cool; when it is cold it can be frozen for later use.

Gomme has not always been cooked in this way everywhere in Norway. Before turning to Setesdal and northern Norway a look at Jensön's Norwegian dictionary (1646) reveals the seventeenth century method of cooking the dish: '*Gumbe*, a kind of cheese, is made from fresh milk, about 3 buckets; it cooks for 1 day and is stirred all during the day; just before it becomes cheese, thin sour cream is poured on as well as egg yolks and melted butter'. He does not say why these latter ingredients are so used, nor whether there is continued cooking of the dish.

Pontoppidan (1749) takes us a step further. He tells about decorating the dish: '*Draule* is pricked (*bestikket*) with raisins and currants, often with pepper.' He also tells that *gomme* from Nordland was sold in Bergen, but does not explain how the *gomme* was preserved for transport over long distances. We can assume, of course, that *gomme* was a commercial item already in his day.

In Setesdal they used fresh milk, either with rennet or home-made sour milk, for making *gomme*. No sugar or eggs were used. According to Gjerden's work on Setesdal foodways, *gomme* made in Bykle, in the northern part of Setesdal, was made from 50 litres of fresh milk heated to *c*. 35°C, and with two tablespoons of rennet added. The preparation was removed from the heat, and the milk and rennet were rested for half an hour until the curds had formed. The mixture was then returned to the heat, stirred gently and brought to the boil, and allowed to boil until only a little whey was left. Care was taken that it was not boiled quickly, so that the dish did not get a dark colour. From 50 litres of milk one got between 25 to 30 litres of *gomme*.

Further south in Setesdal, in Valle, to make real *gomme* one should begin by preparing sour milk, which was done by leaving unskimmed milk to sour. In making *gomme* it was best, it seems, to use a copper pan, newly cleaned and greased with fat while it was cold. In Valle it was usual to make *gomme* from 70 to 80 litres of fresh milk, using the ratio of 1 litre of sour milk to 6 litres of fresh milk, to curdle it. The fresh milk was brought to the boil and stirred at the bottom of the pan all the time. The sour milk was then added and the heat switched off. The mixture was then rested for half an hour. A thick layer of curds gathered at the top. This had to be stirred gently at the bottom and brought to the boil again. There should be big pieces of curds at this stage. The curds were then boiled gently for up to 10 hours, and when ready, the curds and whey had separated into equal proportions. All should have a sweet taste and a yellow/brown colour. The ready dish was then poured into other pots, and the remaining whey was absorbed by the curds. Good *gomme* consisted of big, porous pieces of curds, and little whey.[12]

However, for what the Setesdal-author calls 'false *gomme*'—a variety far less time-consuming to prepare—sugar and eggs are added. It is made from 200 grams of sugar for caramel, three to four litres of whole milk, and approximately one litre of sour milk, plus three to four eggs and thickening. The procedure is almost the same as for *gomme* from Finnøy which is as follows: caramel is prepared, the whole milk is added and brought to the boil. The mixture is stirred gently just before coming to the boil. At this stage the milk takes on a sweet taste and a nice yellow colour. Sour milk and eggs are mixed and stirred into the boiling milk. The preparation is stirred gently and left to rest for a while. During this time the milk curdles. Then the milk must be stirred from the bottom, and it can cook fiercely without a cover until there is about half the portion left in the pot. Then thickening made from 'maizena' (corn flour) and milk is added, and this mixture is brought to the boil. Instead of corn flour thickening one can also use normal thickening consisting of flour and milk, but this has to be allowed to thicken for a while and then is cooked a little after it has been added. This variant is not considered real *gomme* by people in Setesdal but rather an adaptation for use in the home today.[13]

According to Lind Grøttland, the collector of food traditions from South Troms, festive *gomme* needed fresh milk and rennet, and flour for binding the curds. When spices were available, cinnamon, cardamom, and sometimes even bay leaves were used. Preparation of this *gomme* was far more complicated than making the dish in the Setesdal or Finnøy way. Rennet was added to 10 litres of fresh milk. The milk thickened and was then stirred to separate the curds from the whey. The curds (cheese, says

Lind Grøttland) were ladled into a wooden bowl. Another 10 litres of milk were brought to the boil and curdled with buttermilk. The curds from the second batch of milk were placed in the bowl with the curds from the first batch. The whey from the second lot simmered until the desired brown colour was achieved. (Whey from the first portion was fed to calves). Then all the curds from both lots were finely rubbed and added to the whey. This mixture was boiled over low heat and stirred all the time until almost solid. A little of the finest quality flour was sprinkled on it and spices added. The *gomme* was allowed to cool and then it was either rubbed once more, or put through a meat or potato grinder. Finally it was shaped on a flat plate, left overnight, and was then ready for use. This *gomme* could be cut into fine slices and put on buttered bread, waffles and thin pancakes (*lefse*). Lind Grøttland states that curds from milk and rennet were somewhat viscous, while curds from milk and buttermilk were dry, and people in her region preferred to mix these two kinds of curds for *gomme* of perfect quality.[14]

FESTIVE DISH

From as far back as our evidence goes, it is clear that *gomme* has been no ordinary dish. As a festive dish it belongs to the desserts served after the main course of the festive meal. *Gomme* was a festive dish used by the peasants, as the older dictionaries state, instead of cakes (*bakkelse*), but no specific occasion for its use is mentioned in the sources. In Setesdal, however, *gomme* was served at funerals and weddings, but not regularly at Christmas. The large quantities made in Setesdal were needed for the many guests attending a funeral since it was customary to invite family and neighbours to them (the term 'neighbour' is here used in an inclusive sense). The guests usually brought fresh milk to the house of the deceased and *gomme* was cooked there.[15] In contrast to funerals, weddings were attended by fewer guests and thus smaller portions of *gomme* were prepared.[16]

In northern Norway *gomme*, food considered worthy of a lord, was part of the food contributions at weddings, christenings and funerals. It was also used on long journeys and when people rowed to church. When the menfolk departed for the Lofoten fisheries, *gomme* was among the provisions they had with them.[17] *Gomme* was also cooked on returning from the mountain pastures.[18] I came to know *gomme* as a Christmas dish in my mother-in-law's home on Finnøy. I have cooked *gomme* for Christmas every year since 1969 when my own family celebrated Christmas in Bergen and no longer went to Finnøy. Due to modern ways of preservation, we cooked *gomme* in November. Early on Christmas Eve the frozen *gomme* was thawed out in the refrigerator, ready to be served chilled late at night, and to be eaten together with cakes. *Gomme* from Finnøy cannot be sliced—it is not meant to be put on to something else.

Gomme can also, sometimes, be taken as a gift to new mothers. In 1991, on Finnøy, I was served *gomme* at an eighty-fifth birthday celebration—but this is a modern development in the use of this dish.

All three areas mentioned so far also had a lesser quality *gomme* reserved for weekdays which was made from simpler ingredients, mainly skimmed milk. This variety could be used as a main dish, and is sometimes given another name; my family-in-law on Finnøy, and in Stavanger still cooks dravle on cold autumn evenings. On Finnøy, *dravle* differs from *gomme* not in regard to ingredients, but in relation to the time it takes to cook it. In Setesdal the everyday variety is called *haglette*.[19] I have not found different names for the everyday dish in northern Norway; Lind Grøttland only speaks of *gomme* no. 2 when she describes the everyday variety.[20]

GOMME TODAY

I have discussed *gomme* from Finnøy, Upper Setesdal and South Troms. The Finnøy *gomme* presents the modern variety of a dish for festive occasions. I have also pointed out that parallel to the festive dish there exists an everyday variety as well.

Today *gomme* is both a home-made and an industrially produced dish. Vikebygd Dairy, Rogaland, has produced *gomme* all year round for more than twenty years. Ryfylke Dairy on Finnøy has also done so for the last five years. This means that *gomme* can now be had at any time of the year. The old ways of cooking *gomme* were rather labour intensive and the ingredients were expensive. In those days such a dish emphasised the special quality of the occasion on which it was served.

Gomme also expressed the competence of the woman who made it, as a cook. Her efforts were judged by the family, and since *gomme* was also cooked in other houses, by the wider community as well. Thus, *gomme* presented as a food gift (*førningsmat*) indicated a woman's competence as a member of a community, and could ensure her status in a society where social control was strong. Where *gomme* is cooked in, or brought to, another house, community links are strengthened. At the rites of passage, i.e. weddings, christenings and funerals, the dish underlines the importance of these events.

Gomme served as a long journey was about to begin drew the family together, before one or several of its members left with an uncertain return. After the end of the Lofoten fisheries' season the men were received home with newly cooked *gomme*[21] to celebrate the ending of a three months separation. On returning from the mountain pastures, when milk production had reached its peak, *gomme* expressed the joy of this event, of going home after a successful season in the mountains.

Today's higher standard of living has strongly influenced daily life. Families are smaller, people have moved from the countryside to urban centres. Social gatherings such as funerals and weddings often nowadays take place outside the home, and a caterer arranges the food. Birth takes place mostly in hospitals, and although *gomme* is occasionally still brought to the new mother, flowers and baby clothes have largely replaced the traditional food gift.

In addition high quality food is available and consumed throughout the year. The rule that certain dishes should be reserved for particular occasions is no longer as strictly observed as it was in earlier times. Modern living conditions have led to changes in cultural awareness; we no longer necessarily distinguish between food reserved for festive occasions and everyday fare. Formerly milk and milk products belonged to the daily diet, and at the same time, milk and its products, such as butter and cheese, were sold, but when surplus milk was available, milk products were used to make festive dishes, such as *gomme*.

Industrially-produced *gomme* now makes the dish available all year round. However, *gomme* is still cooked at home, and competence is still needed to make a perfect dish. In contrast to the older days, it is cooked, although in smaller quantities, not because of need, but because of pride in regional food traditions. Finnøy people think their *gomme* is the best; theirs is the real thing! *Gomme* from other districts may be tasty—but it is not real *gomme* in their estimation! There is an increasing tendency—linked to regional identity—to take up traditional cooking again. And just because *gomme* is such an adaptable dish, daughters-in-law will continue to learn from their mothers-in-law how to make a particular variety, and their children will go on demanding that it be cooked on special occasions.

NOTES AND REFERENCES

1 J. Brynildsen, *A Dictionary of the English and Dano-Norwegian Languages*, Copenhagen, 1902. Brynildsen gives two meanings for curd: cheese and dish.
2 Gro Nomeland Gjerden, *Gamle mattradisjonar frå Øvre Setesdal*, Bø, 1987.
3 K. L. Grøttland, *Daglig brød og daglig dont. Fra nord-norsk husstell og hjemmeliv*, Oslo/Bergen, 1962.
4 C. Jensön, *Den Norske Dictionarium eller Glosebog*, Copenhagen, 1646. Fasc. With introduction by Per Thorson, Bergen, 1946.
5 K. Leem, *Norske maalsamlingar fra 1740-aari*. (Handskr. nr. 597. 4to i Kallske samling). Utg. ved Torleiv Hannaas. (Skrifter utg. for Kjeldeskriftfondet 48). (Kristiania, 1923), 64, 211.
6 H. Hamre (ed.), *Erik Pontoppidan og hans glossarium Norvagicum*, Bergen/Oslo, 1972. (Årbok for Universitetet i Bergen. Hum. Ser. No. 2), 50. This is the first Norwegian dictionary printed in Norway—in Bergen.
7 I. Aasen, *Norsk ordbog med dansk forklaring*. 3. opplag, Christiania, 1909.
8 *Norsk ordbok* 2, Oslo, 1978.
9 *Nynorskordboka*, Oslo, 1986.
10 *Gumbe* (Setesdal), *gubb, gubb-ost* (Østerdal), *gumme* (Bergens Stift); see A. Torp, *Nynorsk etymologisk ordbok*, Kristiania, 1919, and N. Lid in note. no. 11, following.
11 N. Lid, 'Gumbe, Gumme', *Maal og Minne*, 1925, 110–112.
12 Gjerden (1987), *op. cit.*, 70.

122 Milk and milk products

13 *Ibid.*, 112, 115.
14 Grøttland (1962), *op. cit.*, 68.
15 Gjerden (1987), *op. cit.*, 14.
16 A. Berg and A. Noss, 'Bedlag' in H. Landsverk (ed.), *Gilde og gjestebud* Oslo, 1967, 88–89.
17 Grøttland (1962), *op. cit.*, 67, 133, 145, and Lid, *op. cit.*, (1925), 111.
18 L. Reinton, *Sæterbruke i Noreg-* I Oslo, 1955, 458.
19 Gjerden (1978), *op. cit.*, 71.
20 Grøttland (1962), *op. cit.*, 63.
21 *Ibid.*, 144.

11 The use of whey in Icelandic households

Hallgerður Gísladóttir

I have chosen to describe the use of whey or *mysa* in Icelandic households because in detailing the production and uses of *mysa*, that is the extract left when milk has curdled, one touches on the most important characteristics of Icelandic foodways throughout the centuries.

Icelandic *mysa*, used for several purposes in Icelandic households, comes almost entirely from the production of *skyr*, a special Icelandic dairy product which is made from skimmed milk and rennet. To get the required bacterial growth and consistency, a *þétti*—a few spoonfuls of prepared *skyr*—is needed per barrel. Until about the beginning of this century rennet was made mainly from a newborn calf's stomach. The calf was slaughtered before it had any nourishment other than milk, the stomach was extracted and hung up to dry with the curdled milk still in it. When dry it was placed in a bowl of salt water or whey, and after one or two weeks the extract had become rennet.

In *skyr* production, skimmed milk was boiled to remove undesirable bacteria. The hot milk was then poured into a wooden barrel and cooled to 37°C. Then the *þétti*, and the required amount of rennet, were mixed with some of the warm milk and poured into the barrel while stirring slowly. The barrel was then covered with a blanket so that the milk should not cool too fast. After two hours or so the milk had curdled, and then it was usually sieved, or the whey extract was ladled away.

Old literary and linguistic sources testify to knowledge of a dairy product called *skyr* in all the Nordic countries during the Middle Ages. We will probably never know if this dish was similar to the Icelandic *skyr*, but the method Icelanders use for *skyr*-making has been the same from at least the sixteenth century. It is from that period that we get the first exact description of *skyr*-making. If our Nordic relatives had been making the Icelandic-type *skyr* in the Middle Ages they must somehow have forgotten the method in the course of time. The word *skyr* and related words are in fact known in Denmark, Sweden and Norway, and mean different types of sour milk, but never the Icelandic *skyr*.[1] In Iceland on the other hand, *skyr* became the main dairy product. In the seventeenth and eighteenth centuries, and for most of the nineteenth century, *skyr* made from skimmed milk, and butter made from cream, were almost the only dairy products. The seventeenth and eighteenth centuries were difficult times in Iceland because of the Danish trading monopoly, the cold climate, epidemics, volcanic hazards and famines. Livestock farming decreased so that milk was produced mainly by ewes. Cheese production, which was important in the Middle Ages, almost disappeared, because people considered *skyr* production more substantial. *Skyr* was in fact the everyday dish and was eaten two or three times a day right up to the twentieth century.

The everyday *skyr* was usually mixed with some porridge, often made from Icelandic moss, because formerly there was always a shortage of grain in Iceland. During the first centuries of habitation some corn-farming took place, but after the Middle Ages it ceased, and imported corn was always expensive as it had to be transported from far away countries. Icelanders spared grain, and bread was rare among the common people right up to the last decades of the nineteenth century. It was considered more substantial to make porridge of the little grain that was available, and to use *skyr* to complement it. This shortage of grain characterises Icelandic food habits throughout the centuries, but it also motivated the production of *skyr*, for example. Because of the extensive *skyr* production in the home-farms and summer mountain dairies a considerable amount of *mysa* was made, and every drop was used. Most of it was acidified (*sýra*: 'sour'), and was either drunk mixed with water as *blanda* (blend), or used to conserve food.

The exceptional use of whey in Iceland was early noticed. In *Króka Refs-Saga*, an Icelandic saga believed to have been written in the early fourteenth century, the Norse eleventh century king, Haraldur Harðráði says: 'In Iceland they have a drink called *mysa*. It is all in one: *mysa*, soup and beverage.'[2]

In the adjacent countries corn beer was the everyday drink of the people. Because of the lack of grain, beer never became an everyday drink of Icelandic farmers. However, beer was brewed, especially in the Middle Ages, both from domestically-grown barley and imported malt. But this beer was almost entirely for feasts, and like so many other things, it stopped almost completely during the above-mentioned Dark Ages. This situation promoted the production of sour whey which, mixed with water, was the general beverage in Iceland. It was 'cured' in a special way; it was poured into a wooden barrel, with open holes on the lid, for fermentation. As the whey fermented bits of *skyr*, scum and impurities bubbled up through the holes. These were cleaned away as soon as they appeared and fresh *mysa* was added to keep the barrel full. When the impurities stopped coming to the surface the holes were closed, and the barrel was stored in a cold place for a while. How long the barrels were kept before the whey was considered sour enough depended on the preference of the family concerned. Some thought it was ready for drinking after some months, while others kept it one or two years before using it. When the whey was sour enough, the *mysa* had become *sýra*. The *sýra* was mixed with water as is done with fruit extracts today, and, as already stated, was called *blanda*. It was common to mix eleven parts of water with one part of sour whey or *tólftarblanda*. This was the Icelanders' everyday drink. The taste of sour whey was sometimes sharpened with some herbs or berries; in some parts of south-east Iceland a bag with dried thyme was placed in the whey-barrel to improve the taste.

Sour whey was purchased from the milk districts by the fishing stations. According to law the shipowners had to supply their fishermen with a prescribed amount of *sýra* for the fishing season. Where dairy products were especially difficult to come by, the fishermen made their own artificial *sýra* by souring water with rye or acid herbs.

The vast number of words in the Icelandic language for the different stages of *sýra* production, scum and residue, and its role in Icelandic belief and custom, shows how much this product has been part of the national consciousness; from the seething in the *skyr* and whey-barrels, or from the scum, people could even forecast the weather and mark the tides.[3] Icelanders also transferred words used for beer brewing to the *sýra* production. At the beginning of the twentieth century, when it was forbidden to import and sell alcoholic beverages in Iceland for a time, there were some farmers in almost every district who brewed and distilled alcohol. Then the fermented liquid was often called *mysa* or *blanda* before it was distilled.

But whey was not only used as a beverage; it, rather than salt, was also the most important food conservation medium in old Iceland. In the Middle Ages some salt from sea-water and seaweed was made in Iceland, but domestic production almost disappeared after the sixteenth century, and during the seventeenth and eighteenth centuries, imported salt was too expensive for home use. Therefore, Icelanders dried and smoked fish and meat, and preserved various kinds of food in sour whey. Even the butter was unsalted but sour; foreign tourists writing about Iceland at this time remarked on the sour butter, especially its smell. Meat, fish and even vegetables were put in sour whey. Icelandic women made lots of blood sausages during the slaughtering season in the autumn, and they were all put into the whey-barrels for souring. Liver sausages were extensively produced since the nineteenth century, and these, too, were soured. 'Leathery' meat was cooked and put in the whey-barrels because the acid tenderised it, while better quality meat was smoked or eaten fresh. Animal heads and feet, especially of sheep, were singed, scraped, cooked, de-boned, pressed and soured. Udders, testicles ('spectacle sausages'), flanks, briskets, birds, hardboiled eggs, meat of seals and whales, fat fish, and other kinds of food, floated among the blood sausages, formerly the main food item in the whey-barrels. Throughout the centuries, little grain was used in these sausages, but instead, chopped Icelandic moss was included. In the second half of the nineteenth century when grain became more accessible, women mixed Icelandic moss and rye for the sausages. At the beginning of the twentieth century, the mixture was commonly only one-third moss and two-thirds rye.[4] Gradually the moss element in the sausages diminished and the rye content increased, and during the first decades of the twentieth century, the use of moss in blood sausages stopped. With a good deal of rye in the sausages *sýra* was not so necessary to conserve them because of the acid effect of the rye; it was enough just to pour water over the sausages. The rye acidified the water, and it was possible to put other foodstuffs into the rye-acid in the sausage barrels. This use of rye-acid caused a marked decrease in the old method of acid production in Icelandic homes.

Special dishes were also made from animal bones and *sýra*. Bones of sheep, cattle or fish, were laid in sour whey until the acid had changed them into a pulp. The pulp was cooked until thick and then mixed with *skyr*. *Sýra* was also mixed with a little *skyr* and cooked until it became pulp. These very calcium-rich dishes were always considered the food of the poor and disappeared around the turn of the century. The deposit from the sour-food barrels was also mixed with porridge, and when there was a lack of milk, *blanda* was poured over the porridge instead of milk.

In former times in Iceland, the pantries of Icelandic farms were full of huge barrels which might contain drinking whey, *skyr*, or sour whey with food items for preservation. In excavated medieval farm ruins, the holes for these huge barrels (up to one and a half metres in diameter) are commonly found.[5] They were usually dug well into the earth, and sand or volcanic pumice was placed between the soil and the wood for isolation purposes. A legal text from the early Middle Ages contains a clause requiring tenants who are vacating their leasehold farms to repair the houses if they have had to damage them when removing their barrels.[6] In the old Icelandic sagas there are certain events which also show how huge these barrels could be. Once a chieftain hid in a barrel of sour whey when enemies burnt down his farm. 'This was a small barrel' explains the saga. However, when his enemies came searching for him and perforated the barrel with their spears, they could not find him![7] A twelfth century hagiography of Saint Þorlákur Þórhallson tells of a boy who drowned in a whey-barrel, and who was miraculously brought back to life by invocations to the saint.[8] It was not until the last decades of the nineteenth century that the whey-barrels which were kept in the pantry became smaller in size, and that farmers stopped digging holes for them in the pantry floors.

The barrels used for transportation of the drinking-whey were not as huge as the pantry ones. Sour whey was also transported in containers made from animal skin. A special Icelandic type of wooden stave-vessel commonly with a beautifully carved lid, is called *blöndukönnur* after the *blanda*. These often stood in the doorway of the farmhouses so that everyone, guests or inmates, could have a sip when thirsty. There were also a number of smaller vessels for *sýra* and *blanda*, and these vessels have a wide variety of names as well.

A whey-cheese was also made. Fresh whey was boiled until it became a thick brown pulp, and this was eaten with bread and butter and was considered a delicacy. A lot of fuel was required in order to cook whey-cheese, and as Icelanders were always short of fuel, it did not become all that common until the twentieth century. But in some places there were environmental advantages; at least in the first decades of the twentieth century, people living near the geothermal areas used the boiling spring water for making whey-cheese. The whey was poured into huge shallow iron trays which floated on the hot water while the whey slowly boiled down.

Sýra was used to curdle hot milk for making a cheese called *kjuka*. The curd was sieved, pressed and eaten fresh. Milk porridges and soups were also often curdled with *sýra* and boiled for a while. The most famous of these curds is the Icelandic moss-curd which is still popular in northeast Iceland.

Porridges and soups were made with fresh whey, and fish and meat soups were flavoured with *sýra* or acid *skyr*, but in Icelandic homes, spices other than salt and pepper, were rather rare up to the Second World War, and thus the acid flavour was the most commonly relished one.

Loaves of rye were sometimes cooked in whey. It was rather common to cook grain loaves in liquid—in whey, meat or fish broth—partly because Icelanders did not have any kitchen stoves with ovens until the last decades of the nineteenth century. Before that most of the bread was either unleavened flatcakes baked on slabs on the hearth, or loaves leavened with *sýra*, and baked on the hearth under cauldrons covered with red-hot embers and ashes, or unleavened cakes cooked in whey or broth.

Sýra was not only used for human nutrition in Iceland; it was also considered exceptionally wholesome for young cattle and horses. An old Icelandic proverb says: '*Sýra*-drinking foal will never get any infection.' Livestock were often given *sýra* after human food had been acidified in it. Some farmers gathered bones and cabbage and put them into special animal barrels to use as fodder for their livestock.[9]

Mysa was also used for personal grooming and healing. Women washed their faces with *mysa* for beauty. Very old *sýra* was thought to be good for healing. Poultices heated in such liquid were considered especially soothing for rheumatism, for example.[10]

At the same time as rye-acid production increased, Icelanders stopped weaning and milking ewes, and in the third, fourth and fifth decades of this century, dairy stations were established in various districts. Accordingly, *skyr* production moved from the farms to the dairies, and the old method of making *sýra* disappeared. The *mysa* that was produced in the dairies was considered useless for beverage-making as it tended to putrefy instead of souring as it did formerly. (This situation might have something to do with the sterilisation method used in dairies.). It was mostly used for whey-cheese, and the rest was just poured away or given to livestock. After the middle of this century, the use of whey in Icelandic homes almost disappeared for a while, but in the last two decades, its use has grown markedly again, partly because of the 'health revolution'. *Mysa*, having vitamin B, minerals and little fat, is thought to be an especially healthy food. Lately, the dairies have been promoting *mysa*, and in their popular cookery books it is recommended for several dishes. It is again common to have *mysa* to hand in the home—now in a paper carton in the refrigerator. It actually gives the same taste as white wine when used, for example, in sauces, soups, and sea-food dishes. It is also common once again to cook fish in *mysa*, and people drink it on its own, or mixed with fruit juices and soft drinks, or in cocktails. The dairies have also put *mysa*/fruit beverages on the market, and *mysa* with a higher acid content for food souring.

NOTES

1. Inger Larsson, *Tätmjölk, tätgräs surmjölk och skyr*, Stockholm, 1988, 98.
2. *Íslensk fornrit* XIV, 1959, 154.
3. Unpublished manuscript references from the Ethnological Archive of the National Museum of Iceland are given throughout under the following signa: þþ 368; þþ 391; þþ 396; þþ 1154; þþ 2908; þþ 3962.
4. þþ 7; þþ 14; þþ 75.
5. Sveinbjörn Rafnsson, 'Sámsstaðir í Þjórsárdal'. Árbók Hins íslenska fornleifafélags 1976, 39–120, 105–106.
6. *Grágás—Islændernes lovbok i fristatens tid*, ed. Vilhjálmur Finsen. *Konungsbók I–II, Oslo, 1974, 138*.
7. *Sturlunga saga I*, Reykjavík, 1946, 492–493.
8. *Biskupa sögur I*. Reykjavík, 1948, 170–171.
9. *Guðmundur Þorsteinsson frá Lundi. Horfnir starfshættir*, Reykjavík, 1990, 89; pp 5781; þþ 370; þþ 29; þþ 9567; þþ 9687.
10. þþ 370; þþ 29.

SELECT BIBLIOGRAPHY

K. Eldjárn, 'Aðsauma síl og sía mjólk', *Árbók Hins íslenska fornleifafélags 1960*, 1960, bls. 48–63.

H. Gísladóttir, 'Ostar sáust helst í farteski ferðamanna', *Lesbók Morgunblaðsins 29. okt.*, 1988.

H. Gísladóttir, 'Um mjólk á fyrri tíð', *Matarlyst—120 uppskriftir*, Reykjavík, 1990.

H. Gísladóttir, Eldhús og matur á Íslandi, Card. Mag. thesis, University of Iceland, 1991.

S. V. Guðjónsson, 'Mjólkurmatur', *Manneldi og heilsufar í fornöld*, 1949, 195–211.

E. Hitzler, *Sel—Untersuchungen zur Geschichte des islandischen Sennwesens seit der Landnahmezeit*, 1979.
J. Jakobsson, 'Um mjólkur not á Íslandi', *Rit þess konunglega íslenska loerdómslistafélags* XI, *1791, bls. 193–241*.
B. Johnsen, 'Skyr', *Árbók Landbúnaðarins*, 1990, bls. 71–76.
A. Sigurðardóttir, 'Aðkoma mjólk í mat', *Vinna kvenna á Islandi í 1100 ár*, 1985, 221–264.

12 Koumiss in Mongol culture: past and present

Joyce S. Toomre

Koumiss, the fermented milk of mares, has been an important food product in the steppes of Central Asia since ancient times and is still widely drunk in Mongolia today.[1] The English word comes from the Russian *kumys*, which in turn is based on the Turkic *kumyz*.[2] Outside of Mongolia, koumiss is consumed mainly in Scandinavia and in the former Republics of the Soviet Union, as well as in areas that share borders or traditions with Central Asia. In Norway, koumiss is known as *kaeldermelk*, in Finland as *fuli* or *puma*, in China as *ma tung*, and in the Middle East[3] as *situta*.

KOUMISS, ITS PRODUCTION, COMPONENTS, AND VARIATIONS

The preparation of koumiss has remained unchanged for millennia. According to Herodotus, whose description of the procedure in the fifth century BC is the earliest known, the Scythians produced koumiss by pouring mare's milk into hollow wooden containers, which were then shaken by their slaves.[4] Nearly eighteen hundred years later, Friar William of Rubruck, one of the first modern European visitors to Mongolia, gave further details of the process. Before the milk can be churned, it must first be collected, which is no mean feat since mares are notoriously difficult to milk. Mares living in steppe conditions give about two litres of milk per day and are milked four or five times a day or even more frequently. The job of milking the mares was reserved for the men; the women milked only the cows and the smaller animals. According to Rubruck, when it was time to milk the mares,

> they stretch above the ground a long rope between two stakes stuck in the soil; and . . . tether to the rope the foals of the mares they intend to milk. Then the mares stand beside their foals and let themselves be milked peacefully. In the event of any of them proving intractable, one man takes the foal and puts it underneath her to let it suck a little, and then withdraws it while the milker takes its place. So having collected

a great quantity of milk, which when fresh is as sweet as cow's milk, they pour it into a large skin or bag [*bucellum*], and set about churning it with a club which is made for this purpose, as thick at the lower end as a man's head and hollowed out. As they stir it rapidly, it begins to bubble like new wine and to turn sour or ferment, and they keep churning it until they extract the butter.[5]

The koumiss (also called *airag, arrag* or *chige*) required considerable stirring and was drunk only after it had become fairly pungent. Upon first tasting koumiss, Friar William admitted that he 'broke out in a sweat all over from alarm and surprise, since I had never drunk it before. But for all that I found it very palatable, as indeed it is'. Elsewhere he noted that 'while one is drinking it, it bites the tongue like *râpé* wine, but after one has finished drinking it leaves on the tongue a taste of milk of almonds. It produces a very agreeable sensation inside and even intoxicates those with no strong head; it also markedly brings on urination'.[6]

Unfermented mare's milk is rarely drunk as it is a laxative and strong purgative. Once fermented, however, mare's milk is easily digested. Koumiss nowadays is prepared with two starter cultures, *Streptococcus lactis* and *Lactobacillus bulgaricus*. Aside from producing lactic acid, the *Lactobacillus bulgaricus* also produces acetaldehyde, which is important for the final flavor.[7] Depending on the length of fermentation, the finished product contains 2 to 2.5 per cent protein, 1 to 2 per cent fat, 3.5 to 4.8 per cent milk sugar, 0.6 to 1.2 per cent lactic acid, and 1.0 to 3.0 per cent alcohol.[8] A variation of koumiss is sometimes made from donkey's milk. Similar fermented drinks are *kefir*, made from cow's milk, and *shubat*, made from camel's milk. Koumiss when distilled is called *arkhi*. It is unclear exactly when and from whom the Mongolians learned the art of distilling, which was apparently not known during Rubruck's visit, but was known a hundred years later.[9] A refined version of koumiss reserved for the noblemen was *caracosmos (qara qumis)* or black koumiss. It is no longer made, but in Rubruck's time, the milk of black mares (*qara* is Turkish for black) was collected and beaten until all the solid particles sank to the bottom, like the dregs of wine. The liquid on top was clear and sweet unlike ordinary koumiss, 'which was white, cloudy and sour-tasting'.[10]

RITUAL USE OF KOUMISS IN TRADITIONAL MONGOLIAN CULTURE

Mongolia in the early thirteenth century was a land of horsemen and warriors ruled by the legendary Genghis Khan (*c.* 1162–1227). Part of the conqueror's success was due to his warriors who, with their steeds, were fleet of foot and slow to hunger. They carried little with them on their campaigns except pieces of dried meat hung from the saddle and bags that were filled with dried milk, curds, and koumiss. In this society everything belonged to the Emperor. All males, even the chieftains, were 'obliged to give mares to the Emperor as rent so he could have their milk for one, two or even three years'.[11]

The traditional Mongol diet consisted primarily of dairy products or meat, with beef or mutton preferred over goat or camel meat. White foods came in many varieties and, in late spring, they were often the only foods. Slaughtering practically stopped at this season because the animals were bearing their young which were needed to replenish the herds for the next year. When fresh meat was available, it was roasted or boiled and usually was eaten unseasoned save for salt; onions and garlic were known and consumed, but apparently were not used to enhance the flavors of broths and stews. Although the boiling of meat was common, there was rarely much of it. According to Rubruck:

> With the meat of a single sheep they feed fifty or a hundred men: they cut it up into tiny pieces on a dish along with salt and water (since they make no other sauce); and then on the end of a knife or a fork made specially for the purpose—the sort with which we usually eat pears and apples baked in wine—they offer each of the bystanders one or two mouthfuls, depending on the numbers at the meal.[12]

Horseflesh, and fish, were eaten in the early period, but the consumption of both was outlawed by the Third Dalai Lama upon the conversion of the Mongols to Tibetan Buddhism in the latter part of the sixteenth century.[13] Grain was a scarce commodity for the nomads and obtained, not by farming, but by trading or plundering. Bread was not part of the diet, but they boiled millet into a soupy kind of porridge and used what wheat they had for noodles.[14] By the sixteenth century, agriculture was well established, at least in the border areas near China. According to the Chinese border expert, Hsiao Ta-heng, writing in 1594, the Mongol nomads 'grow wheat, millet, beans and glutinous millet. These cereals they have cultivated for several generations—it is not something which has started recently'.[15] In Rubruck's time, the Mongols were not yet drinking tea but, by the fifteenth century, the aristocrats were drinking it, and after the 1570s the custom of tea-drinking had become widespread even among the common people. Ultimately the Mongols adopted the Tibetan style of preparation whereby salt, milk or cream, and roasted millet were added to the boiling tea.[16]

Of all the white foods, koumiss was the favourite. The chieftain's offer of koumiss to a guest was considered an honour, and koumiss played a part in most of the traditional Mongolian rituals. Upon entering a Mongol tent, the visitor faced the master and mistress who sat in places of honour opposite the entrance. Over the master's head, fastened on the wall, was a little felt idol or doll; a similar one hung over the head of the mistress. Another smaller image, hung higher up the wall between the two, was the guardian of the entire house. 'Close to the entrance on the women's side is yet another effigy with a cow's udder for the women who milk the cows, since milking the cows is one of the women's tasks: on the other side of the entrance, facing the men, is a second statuette with a mare's udder, for the men who milk the mares'.[17] Before drinking, some of the beverage was sprinkled first on the master's effigy and then on the others in turn. Then a steward took the goblet and went outside the tent where he sprinkled some of the beverage at each of the four points of the compass. Before the master drank, he poured some on the ground or, if he was mounted, he poured some on the horse's mane. When the idols needed renewing, the chief ladies of the clan gathered together to make them; when the job was done, they killed a sheep, ate it, and burned the bones in the fire. Carpini, another early visitor from the West, noted that the Mongols also offered the first milk of every cow and mare to their idols.[18]

A table or bench, often ornamentally carved, painted, or inlaid with bone or silver, invariably stood near the door just inside the tent or yurt. Goblets and cups were arranged on the table along with containers holding koumiss and sometimes other beverages like wine, rice wine (*terracina*), or honey mead (*bal*).[19] It was considered a polite gesture on the part of a visitor entering the tent to stir the koumiss as the stirring aided the process of fermentation. The vessels holding the koumiss were called *saba*; they were made of animal skins and could be enormous in size. Those of wealthy Kazakhs had a capacity of nearly 500 litres and took up to ten horse hides to produce.[20] The wooden ladles for stirring the koumiss were often carved in the form of two joined scoops. Scholars do not know the origin of this tradition, but it has been suggested that 'the ancients used one half of the scoop to pour kumys dedicated to the sky spirits and the other for those of the underworld'.[21] At the entrance of the Emperor's tent or of one belonging to a Tartar prince, one or more musicians with their instruments stood by the table with its sack of koumiss, ready to play the moment the master began to drink. At large feasts all the attendants danced and clapped their hands, the men before the master, the women before the mistress, as long as the drinking lasted. When the drinking stopped so did the music.[22]

The first appearance of the year's new koumiss was celebrated by a great festival on the ninth day of May. This was a joyous occasion for the Mongols, but before the drinking of the koumiss began, the Mangu Chang's soothsayers first consecrated all the white mares in the herd and then sprinkled some newly-made koumiss on the ground. According to Rubruck, this was 'like the custom that obtains among us in some places regarding wine on the feast of St Bartholomew or St Sixtus, or with fruit on the feast of Saints James and Christopher'.[23] Two great drinking sessions were held each year at the Mangu Chan's palace in Caracorum, once when he passed through the area with his retinue and herds about Easter, and again in the summer on his return. This palace was so grand that the usual horsehide containers for koumiss and other drinks were replaced by a large silver fountain, shaped like a tree. This bizarre device was commissioned by the Mangu Chan from Master William of Paris, a goldsmith who lived among the Mongols. Rubruck's description of the fountain, although lengthy, vividly conveys the sense of the court's opulence and of the festive role of koumiss and other drinks in the society:

> [There were four silver lions at the root of the tree], each one containing a conduit-pipe and spewing forth white mare's milk. There are four conduits leading into the tree, right to the top, with their ends curving downwards, and over each of them lies a gilded serpent with its tail twined around the trunk of the tree. One of the pipes discharges wine, a second *caracosmos* (refined mare's milk), a third *boal* (a drink made from honey), and a fourth rice ale, known as *terracina*. Each beverage has its own silver vessel at the foot of the tree, ready to receive it. Between the four pipes, at the top, he made an angel holding a trumpet, and beneath the tree a cavity capable of concealing a man; and there is a pipe leading up to the angel through the very core of the tree ...
>
> So when drink is required, the head butler calls to the angel to sound the trumpet. On hearing this, the man concealed in the cavity then blows strongly on the pipe that leads to the angel, the angel puts the trumpet to its mouth, and the trumpet gives out a very loud blast. When the stewards in the chamber hear this, each pours his drink into the appropriate pipe, and the pipes spurt it out, down into the vessels designed for the purpose; whereupon the butlers draw it up and convey it through the palace to the men and women.[24]

To be without koumiss was a hardship in this society. More than that, the absence of koumiss threatened a man's ability to survive in the desert where food and drink were always in very short supply. Rubruck told of a certain Saracen who wanted to be baptized, but changed his mind after his companions told him that Christians were forbidden to drink koumiss. This notion was seemingly spread by the Russians in the area and was responsible for alienating many Saracens from the Christian faith. Although Rubruck himself drank and enjoyed koumiss, he could not convince the man that the adoption of Christian faith would not mean the proscription of his beloved koumiss.[25]

Koumiss was so important to the Mongolians, that when a nobleman died, he was buried in one of his yurts, seated in the middle of it before a table on which was placed 'a dish filled with meat and a goblet of mare's milk'. Also buried with him were a mare and her colt, and a stallion with bridle and saddle, along with other valuable possessions of silver and gold.[26] Hsiao Ta-heng, writing at the end of the sixteenth century, noted that armour and clothing as well as concubines, secondary wives, and servants were buried with Mongol nobles and khans in ancient times, but that this practice died out with the adoption of Tibetan Buddhism.[27]

KOUMISS IN MODERN MONGOLIAN CULTURE

Fermented mare's milk has continued to play an important part in nomadic culture, right up to the present day. Over the years, travellers in Mongolia and Central Asia repeatedly described the importance of koumiss to the local population. These descriptions circulated fairly widely and even turned up in some unlikely places. For instance, in the chapter on cheese in the nineteenth century French book of cookery and estate management, *Maison Rustique*, the author, a certain Comtesse de Genlis, reiterated that koumiss was so much to the taste of the Russian Tatars that 'all their happiness consists in always having a large quantity of it available'.[28] The food historian, V. V. Pokhlebkin, in writing of the cuisines of the, now former, Republics of the Soviet Union, mentions the importance of soured milk products, including koumiss, in the diet not only of Mongolians, but also of the Kazakhs and Kirghiz. Aside from koumiss, the Mongolians, in particular, are noted for the sheer variety of their milk products, with many prepared with a mixture of three to five kinds of milk, including the milk from mares, cows, sheep, camels and yaks. For the Kazakhs, typical festive dinners begin and end with the drinking of koumiss and tea. Pokhlebkin also includes a recipe for a traditional Kazakh thick soup, *Nauryz-kozhe*, that is made from several kinds of milk, including koumiss. The milks are mixed with several varieties of meat, including fresh horse or camel flesh, thickened with wheat or rice, and seasoned with onions, garlic, and red pepper.[29]

Outside of ordinary consumption and use in the kitchen, koumiss continues to be ritually significant in modern Mongolian culture. For instance, koumiss 'the best tasting of food and drink' is an important motif in the stylized poems that are exchanged between the families of the bride and groom on the wedding day.[30] Koumiss is also important in the annual cycle of festivals which follow the lunar calendar. The most important festival occurs on what Westerner's call New Year's Day, but what for Mongolians is the beginning of the 'rich white month'. On this day the family gathers outside the yurt at dawn and bows first to the sun rising in the East, and then to the other points of the compass. They then make a sprinkled offering to heaven using mare's milk, cow's milk, or even wine. Later, following homage to the family's Buddhist images inside the yurt, the eldest son customarily presents a bowl of koumiss to his parents, but 'before receiving the bowl, the father dips his middle finger into the kumiss and sprinkles it towards heaven three times as an offering to the gods'.[31]

May is still the time for the spring festival and the first milking of the mares to make koumiss. This tradition has continued uninterrupted through the centuries. Like Rubruck before him, the eighteenth century German naturalist, Pallas, described the consecration of the mares and the drinking of koumiss at the spring festival.[32] For many Mongolians today spring, with its luxurious vegetation, remains the favourite season. It is a time, not only to make fresh koumiss, but also to examine the herds and to decide which stallions to keep for breeding, and which to castrate. 'Because many of the horses have never been ridden or trained, spring is also the time when young Mongol men show their prowess and courage in saddling, breaking, and training horses'.[33]

Over the summer, many *oboo* festivals are held throughout Mongolia. These festivals are combinations of administrative meetings, religious ceremonies, sports events, and social gatherings. Political affairs are decided, deities are honored, and young men display their skills in wrestling, archery and horse racing. The offering and drinking of koumiss remains an important part of the festivities, just as it has for hundreds of years. Women are excluded from the religious ceremonies, where the chief male dignitaries, including the ranking lama, 'gather to offer kumiss and make sacrificial burnt offerings'. Refreshments, including koumiss, are then served in a tent erected for that purpose. This is a very formal occasion redolent of earlier imperial court parties in its splendor and brilliance. Again it is a 'male-only' event, with women relegated to the sidelines and to lesser tents. In the main tent is a raised dais for the host, usually a prince or noble. Rugs and felts are spread out on either side of the dais; the religious dignitaries are seated to the right, secular male dignitaries to the left. Opposite the dais and across the tent at the end of a wide passageway is a large container of koumiss and a formally-dressed official, who supervises the distribution of the koumiss to the dignitaries. The main tournament events take place in front of this tent.[34]

The 'month of the rich autumn,' the eighth month of the year, is another time for celebration and the drinking of koumiss in Mongolia. At this season, the animals are in good condition and there is a plethora of milk products. This festival which goes back to ancient times is essentially a ceremony of thanksgiving to the gods for the bounty of the flocks, and a beseeching for these blessings to continue. Outside of the yurt and to the accompaniment of chanted sutras, the father of the family waves an arrow with flowing colored streamers in a wide clockwise circle. As he repeats the words '*khurai-khurai* (gather together! gather together!), the cattle are driven around the yurt following the direction of the arrow. After the ceremony, family and friends gather for a party, the highlight of which is the drinking of koumiss.[35] With the onset of winter, the consumption of koumiss is considerably restricted, both for natural and economic reasons. In the old days, only the nobles had enough koumiss to last them over the winter, but today herding is not as widespread among Mongolians and the by-products of the traditional culture are insufficient to meet the demands of the growing urban population. Although the taste for koumiss remains, increasingly the serving of this traditional beverage is reserved for ritual occasions and special events.

NOTES AND REFERENCES

1 A brief word needs to be said concerning the sources for this paper and its methodology. Culinary history is still a new field and is multi-disciplinary at its very core. While it is a disadvantage to be at the mercy of experts without necessarily being able to interpret conflicting opinions, the advantage is the possibility of assembling diverse shreds of evidence to build a composite new picture. In the present case, I have leaned heavily on two texts to prepare this paper: for the traditional use of koumiss in Mongolian culture I have used the Hakluyt Society's edition of William of Rubruck's journey to Mongolia in the thirteenth century (*The Mission of Friar William of Rubruck*. Transl. by Peter Jackson, London, 1990) and for the modern story, Sechin Jagchid and Paul Heyer's important text, *Mongolia's Culture and Society* (Boulder, Colorado, 1979). Without the work of these historians and scores of other Mongolian specialists, I could not have reconstructed the story of koumiss and reformulated it for another audience; to these primary researchers, I extend my thanks.
2 L. V. Clark, 'The Turkic and Mongol Words in William of Rubruck's *Journey* (1253–1255), *Journal of the American Oriental Society*, XCIII (1973), 189. Max Vasmer, the distinguished German philologist, noted that koumiss was a *Tatar* word that ultimately derived from *kummak* (to shake or stir). (Max Vasmer, *Etimologicheskij slovar' russgogo jazyka* Moscow, 1964, Vol. 2, 416–417.
3 G. Campbell-Platt, *Fermented Foods of the World: A Dictionary and Guide* London, 1987, 108. According to *The Great Russian Encyclopedia*, Vol. 13 New York, 1976, 454, koumiss is also widely produced in Latin America, but I have no information to corrobrate that point.
4 Herodotus, *The History*, Transl. by David Grene, Chicago, 1987, Book Four, No. 2, 279.
5 *The Mission of William of Rubruck*, 81
6 *Ibid.*, 99, 81. The editors define *râpé* wine, as an 'inferior wine obtained either by the use of unripe grapes or by adding water to the marc.'
7 Campbell-Platt, *op. cit.*, 109.
8 *Great Soviet Encyclopedia*, Vol. 13, 454.
9 T. Takamiya, *Beiträge zur Geschichte der Nahrung und der Nahrungsbereitung bei den Hirtenvölkern Mittel- und Innerasiens* (Inaugural Dissertation at the Ludwig-Maximilian University, Munich, 1978), 124–129. Most scholars agree, however, that distilling was introduced into Mongolia from the West and not from China. Paul Pelliot noted that the occidental name for the distilled beverage *araki*, or *arkhi* shows that it could not have been known before the thirteenth or fourteenth centuries (P. Pelliot, 'A propos des Comans, '*Journal Asiatique*, 11e série, XV (1920), 170). Michel Hoang, claims that *arkhi* 'originated in India and entered Mongolia via Turkestan by the same route as did Buddhism, and at the same time,' with mention of the preparation first occurring 'in the second half of the thirteenth century'. Unfortunately, however, he does not cite his sources. M. Hoang, *Genghis Khan* New York, 1990, 64.
10 According to HsüTing, a Chinese observer of the Mongols in the Southern Sung period (1127–1276), as reported in *The Mission of Friar William of Rubruck*, 82n.
11 John of Plano Carpini in 'History of the Mongols' in C. Dawson (ed.), *The Mongol Mission*, London, 1955, 28. Hereafter, referred to as Dawson.
12 *The Mission of William of Rubruck*, 79. This quote is an important bit of information that supports the general consensus among scholars that table

138 Milk and milk products

forks were introduced into Europe from Byzantium. At the same time that forks were commonly used in Mongolia, to judge by Rubruck's testimony, they were rare and luxurious items in Europe. In fact, table forks were not much used in medieval Europe, except for sticky desserts; by the sixteenth and seventeenth centuries, even that limited use disappeared in France and England. For more details on the adoption of the table fork in medieval Europe, see B. A. Henisch, *Fast and Feast: Food in Medieval Society*, University Park, Pennsylvania State University Press, 1976, 184–189.

13 Jagchid and Hyer, *op. cit.*, 41.
14 T. Takamiya, *op. cit.*, 32–34.
15 Quoted in C. R. Bawden, *The Modern History of Mongolia*, London, 1968,
16 Jagchid and Hyer, *op. cit.*, 44.
17 *The Mission of William of Rubruck*, 75.
18 Dawson, *op. cit.*, 9.
19 *The Mission of William of Rubruck*, 178.
20 V.N. Basilov, V. P. D'yakonova, V. I. D'yachenko, and V. P Kurylëv, 'Household Furnishings and Utensils' in *Nomads of Eurasia*, ed. V. N. Basilov, transl. by M. Fleming Zirin, Seattle: University of Washington Press with the Natural History Museum of Los Angeles County, 1989, 134.
21 *Ibid.*, 132.
22 *The Mission of William of Rubruck*, 77. The same practice was mentioned by Carpini, who said that no Tartar prince 'ever drinks, especially in public, without there being singing and guitar-playing for them' (Dawson, *op. cit.*, 57).
23 *The Mission of William of Rubruck*, 242.
24 *Ibid.*, 209–210.
25 *Ibid.*, 104. The history of koumiss in Russian culture is a separate, but intriguing topic. Koumiss was first mentioned in the Russian Hypatian Chronicle of 1185 as a drink of the Polovtsians (Vasmer, *op. cit.*, 416). We next have Rubruck's testimony that the Russians in Mongolia considered the drinking of koumiss as incompatible with Christianity. This view apparently endured and was corroborated by Samuel Collins, a Scotsman, who was physician to Tsar Alexis in the seventeenth century. Collins noted that Orthodox Russians of that period considered the eating of horseflesh or the drinking of mare's or ass's milk as *pogano* (unclean) and the consumption of any of these products at any time was forbidden by the Church. (S. Collins, *The Present State of Russia* London, 1671, 13–14). Yet by the late eighteenth century, S. T. Aksakov was writing about the benefits of the koumiss cure for those who suffered from tuberculosis. The koumiss cure became very popular in Russia at the end of the nineteenth century and remains so to a limited extent today (See *Kumys i shubat*, 3rd ed. Alma-Ata, 1979, 67).
26 Dawson, *op. cit.*, pp. 12–13.
27 Jagchid and Hyer, *op. cit.*, 103.
28 Comtesse Stephanie de Genlis, *Maison Rustique* Paris, 1810, Vol. 2, 39. I do not know the source of the author's information about koumiss. The Comtesse de Genlis was a colorful figure. Although she was well-connected in court circles and knew such people as Rousseau and Voltaire, she herself never travelled east of what were then the German states and principalities. I am grateful to Barbara Ketcham Wheaton for bringing this reference to my attention and providing information about her.
29 V. V. *Pokhlebkin, Natsional'nye kukhni nashikh narodov* (National cuisines of our peoples), 2nd ed., Moscow, 1990, 563, 355–363. In the late 1970s, the Kazakhs produced most of the koumiss in the Soviet Union (about 24 thousand tons annually), although the Soviet central planners hoped to increase that number fourfold in Kazakhstan and ultimately to raise the total national production to 150 tons annually. Beyond the local demand, some of this output was used as

part of the koumiss cure at various Soviet mountain sanatoria constructed in the Volga region and Kazakhstan to combat tuberculosis and other pulmonary infections. (*Kumys i shubat*, 4–5; *Great Soviet Encyclopedia*, Vol. 13, 454).
30 Jagchid and Hyer, *op. cit.*, p. 86.
31 *Ibid.*, 116.
32 Quoted in Basilov, *etal.*, *op. cit.*, 135.
33 Jagchid and Hyer, *op. cit.*, 120.
34 *Ibid.*, 121–123.
35 *Ibid.*, 128

13 A Swedish beer milk shake

Nils-Arvid Bringéus

In John Steinbeck's *Cannery Row* (New York, 1945, ch.17) there is a marine biologist, known simply as Doc, whose fondness for beer leads a friend to say: 'You love beer so much. I'll bet some day you'll go in and order a beer milk shake.' The idea intrigues Doc, who wonders how it will taste, whether the milk will curdle, and whether the mixture should be sugared. One day he decides to try it, so he chooses a roadside stand in a town where he is unknown and orders a beer milk shake. When the waitress reacts with disbelief, Doc's lying explanation is that he is under doctor's orders to drink it for his bladder complaint. He instructs her to pour in the milk (without sugar) and add half a bottle of beer. To his relief, the mixture is not so bad: 'it just tasted like stale beer and milk'.

The point of this episode is that beer mixed with milk is supposed to be not only unpalatable, but also incongruous—Doc compares it with 'shrimp ice cream'. Yet mixtures of beer and milk—both cold and hot—are far from new, and have indeed been recommended for their medicinal value. One form called *posset*, is defined by the *Oxford English Dictionary* as 'A drink composed of hot milk curdled with ale, wine, or other liquor, often with sugar, spices, or other ingredients; formerly much used as a delicacy, and as a remedy for colds etc.'.

The present essay concerns the position of similar mixtures in the traditional diet of the southernmost provinces of Sweden, especially the former Danish possessions of Skåne (Scania), Halland, and Blekinge. The beverages concerned consist of milk mixed with beer, particularly small beer, a low-alcohol malt drink.

SMALL BEER

An indication of the importance of small beer in Skåne is evident from its name—*dricka*, which means simply 'drink', or *svagdricka* 'weak drink'. Small beer was therefore *the* drink. 'Small beer and bread keep body and soul together' is an old saying from Skåne, and these were in fact two of the major elements in the Scanian diet; they accompanied every meal. The tankard or pitcher of small beer was always on the table, even between meals, so that anyone could slake his thirst when necessary. Many meals consisted simply of coarse rye-bread and small beer, and this combination was often the best they could manage when they had run out of other food. During the hard work of haymaking or reaping, the pitcher of small beer was taken out to the fields.

People today probably think of small beer as something that was necessary to wash down the dry bread. In former days, however, bread was not always eaten dry, and the small beer was not always merely a drink. The bread could be sopped in the small beer and eaten with a spoon. Small beer was thus food for supping, like sour milk or soup. In this form it could be served hot as *drickasupa* 'small beer sup', or cold as *drickablandning* 'small beer mixture'.

THE USE AND DISTRIBUTION OF COLD SMALL BEER AND MILK

Small beer mixture consists of roughly equal parts of cold milk and small beer. The small beer was originally home-brewed. Later it was bought from a man who drove out from the brewery with a horse and wagon, with wooden kegs containing between ten and fifty litres. In the towns small beer could be bought in every grocer's shop, where it was dispensed from a large cask. Small beer mixture was especially tasty if sweet milk was used, but skimmed milk was also common. On the other hand, there are no recent records of the use of sour milk for the mixture, although this was apparently not unusual in the old days.

As a drink with meals, small beer mixture was one of several alternatives, which included sweet milk and sour milk. It was served primarily at breakfast and supper, but could also be used at dinner, especially in combination with salty food. One informant recalls: 'I remember in particular the dinners for the 'thrashin' folk', the people who helped with the threshing. Everybody sat down around the table in the room, the sowl,[1] bread, and potatoes were passed round, and in the middle of the table was a big dish or bowl with small beer and milk out of which they each took a sup with their horn spoons now and then.' The mixture was served 'both after the main course at dinner (often of fried bacon and potatoes) and for supper. It was kept in a big bowl in the middle of the table, and the family ladled it into their mouths with horn spoons and ate bread along with it'. The mixture was also served with fried potatoes. Usually each person had his own easily identified horn spoon. They all ate from the same bowl, and the frying pan with the potatoes stood in the centre of the table.

Small beer mixture has been used in living memory in the same way as that we find recorded in older sources. In the Helsingborg region A. G. Barchaeus made the following note in his travel journal in 1773: 'They have an evening dish in Skåne which is unknown in the north of the country, viz., sour milk mixed with small beer'. From a description of the deanery of Laholm in southern Halland from the late eighteenth century, we learn that 'breakfast consists mostly of herring and small beer or milk, or these two mixed together'. The author adds: 'In the summer, instead of small beer, some people mix this with milk that has begun to sour'. From western Blekinge, Dean Öller recorded in 1800 that 'for breakfast, which is eaten at eight or nine o'clock in the morning, they serve salt herring or boiled herring, and what they call *syba*, i.e. supping food consisting of a mixture of unboiled milk and small beer'. In south-eastern Skåne breakfast at the beginning of the nineteenth century consisted of 'herring, potatoes, and bread, boiled small beer or, when milk is in good supply, a mixture of milk and small beer'. In a description of Lomma in Bara Hundred from 1828, we read that 'supper never consists of anything but cold milk, or small beer, or a mixture of the two, or else of porridge'. Other accounts from the same hundred state that breakfast consisted of 'herring, small beer, and bread, or milk and small beer mixed and called *drickablandning*'. According to Eva Wigström, the mixture in Rönneberg Hundred in western Skåne in the 1840s, consisted of 'sour milk and small beer', served in earthenware dishes and eaten with horn spoons for breakfast. Supper too usually consisted of 'sour milk mixed with small beer'. On the Landskrona plain in the nineteenth century, the breakfast and supper of salted herring and potatoes was washed down with a mixture of 'small beer and soured milk'. In Göinge the mixture was drunk for both breakfast and the midday meal at the end of the nineteenth century. Breakfast consisted of fried or mashed potatoes. The frying pan was set in the middle of the table, and when people ate they took 'a spoonful of potatoes from the pan and then a spoonful of milk from the glazed earthenware dish that stood beside it. Everyone ate from the same pan and the same dish. Sometimes, if milk was in short supply, they mixed it half and half with small beer. Whichever of the two it was, it generally went under the name of "wet"'.

A somewhat later stage is represented by reports that each person was allowed to mix his small beer and milk in the desired proportions. One informant writes: 'Mother used to have a porcelain jug with sweet milk, by which I mean unskimmed milk, and another jug with small beer on the table. So we could each mix our own'. From a teacher's home we are told that 'with dinner we drank only small beer. But for supper there was always a jug of milk and a bottle of small beer. Each person mixed them to taste in his own glass'. The change in eating habits is described by another informant: 'In the beginning, as I remember, we mixed milk with small beer in an earthenware dish which was put in the middle of the table beside a frying pan with fried potatoes'. This was a frequent breakfast dish, eaten with a spoon by the family members assembled round the table. Later the shared dish and frying pan were replaced by a separate plate for each person, and a coffee cup or a glass, and milk and small beer were served in separate jugs, so the people at the table had to mix for themselves.

The pitcher of small beer which stood on the table even between meals in peasant homes in Skåne, could contain either small beer or a mixture of this and milk. This was also appreciated as a refreshing drink for outdoor work. 'During the haymaking my parents always brought this drink with them to the fields', according to an account from central Skåne, while an informant from the Hässleholm district writes: 'When the farmhands had to go to the outlying lands in the summer for a whole day's work, they always had to have small beer mixture with them in a stone jar. This slaked their thirst better than either water or milk or small beer alone'. This practice evidently comes from a time before it was customary to take along coffee for outdoor work.

Along with its function as a drink with meals and a refreshment, small beer mixture was a special part of the diet of nursing mothers, since it was thought to be good for stimulating lactation. Over twenty informants emphasize this, although sometimes with the reservation that they did not know whether the belief was well-grounded or not. The mixture was drunk not only by mothers who gave birth at home, but was also prescribed in the 1930s in the maternity wards in Lund, Hörby, and Kristianstad. In many cases, it was in the maternity wards of Skåne, that women from elsewhere in Sweden had their first experience of small beer mixture.

The mixture was easy to imbibe and digest. 'Towards the end of one's life, when it was sometimes impossible to eat or swallow any other food, a cheese sandwich with small beer mixture always did them good. I remember that both my mother and grandmother, when they were old, would not have any other drink, except coffee of course'. Another informant mentions that many years before he had 'heard tell of an old man who in his last years could not eat anything but small beer mixture'.

Using the responses to an inquiry which I broadcast on a radio programme in 1970, I have mapped the distribution of the custom. Fig. 13.1 marks the places where the respondents came into contact with small beer mixture, regardless of where they themselves lived. The distribution encompasses Skåne and southern Halland, along with scattered instances in Blekinge and southern Småland. The number of recorded instances per town/parish is indicated by the size of the circles. In view of the distribution of population, it is natural that the large towns are best represented. There are more instances in Helsingborg than in Malmö. It is uncertain whether this is due to chance, or a reflection, of an actual difference in frequency. The rural responses are fairly evenly distributed. The scattered instances from Blekinge and Småland show how the custom became rarer towards the edge of the distribution area.

144   *Milk and milk products*

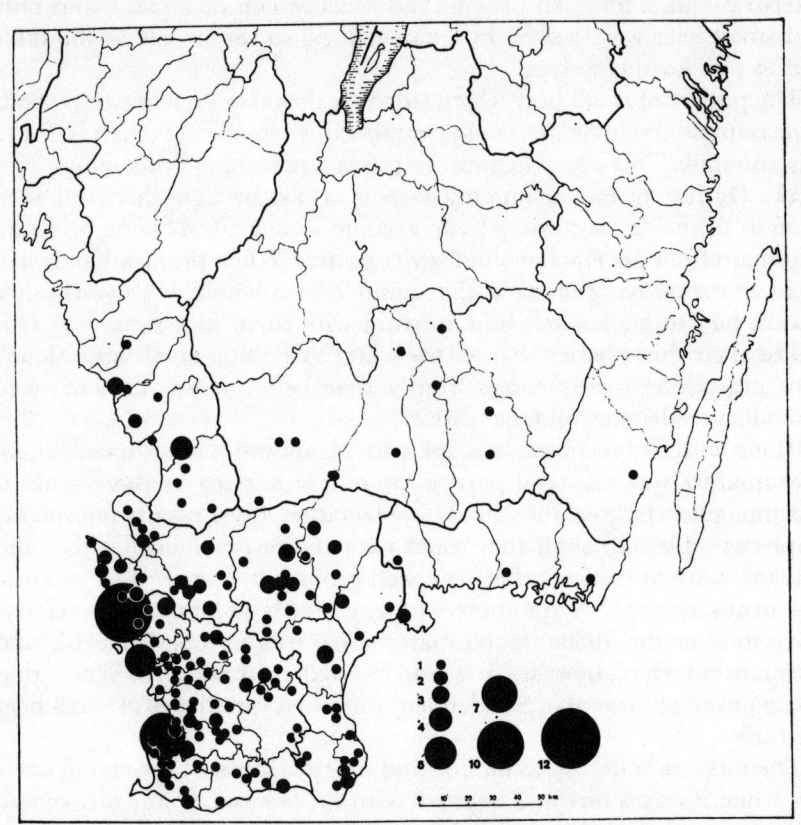

Fig. 13.1. Distribution of small beer mixture according to responses from radio listeners in 1970. (From: *Gastronomisk kalender*, 1972, 142, Stockholm)

The mixture played a very important role in Skåne and southern Halland as late as the start of the present century, in all social classes, particularly in the countryside, but also in the towns. One informant calls it 'Skåne's number one national drink'. Other responses state that it was 'frequent', 'very common', 'regular', or served 'with practically every meal'. In poor homes in particular, the mixture was very important. An informant from the mining town of Billesholm writes: 'Thirty years ago you could rightly say that most working-class children were reared on coarse rye-bread and small beer mixture. And I dare say that in nine out of ten working-class homes, at least here, it was on the table every day. All the miners, with few exceptions, had half a litre of it with them for breakfast'. In southern Halland too, where we have seen early references to the mixture, it is said to have been common, as it also was in the Markaryd district of south-west Småland. Further away from the central area it appears to have been less general. An account from Småland describes it as 'a drink given to servant folk; while from eastern Blekinge we learn that it was found 'in humble and poor homes'. It thus appears as if the mixture was lower down the social scale in the border regions than in the central area, where we know that it was drunk in the homes of merchants and members of parliament.

HOT SMALL BEER WITH MILK AND SIMILAR DISHES

Many people associate the cold *drickablandning* with the hot *drickasupa*. The latter corresponds roughly to what is known as *ölsupa* 'beer posset' in standard Swedish, but it should rather be defined as the hot equivalent of the mixture of cold small beer and milk. Dean Öller reports from 1800 that 'for supper they once again cook hot food, consisting of porridge and milk, or small beer instead of the milk, or boiled potatoes and hot small beer with a little milk in it'. This mixture had the same distribution as the cold mixture. In Denmark it is known simply as *søbe* 'a sup'. This has occurred in living memory in Bornholm, Møn, Lolland-Falster, and was formerly known in parts of Sjælland. It was a common dish in Bornholm.

In Skåne a normal way to use up hard crusts of bread was to make small beer sup at the end of the week, but this was a proper meal, not just a mixture of leftovers like *bänkvälling* 'tabletop gruel'. A record from Harjager Hundred from 1828 states that breakfast consisted of 'herring and potatoes with milk or small beer sup'. This could also be served in a grander variant with sugar, cream, a whisked egg yolk, and spices. In south-west Skåne this mixture was known as *gubbaknorr* 'old man's zest', but could also be called *kvingedricka* 'woman's small beer' since it was given to women in childbirth.

Where milk was not available, it was possible to pour boiled small beer on to the sops. Breakfast in the fishing hamlet of Kivik consisted of salt herring and potatoes, which they 'washed down with small beer and bread'. Supper consisted of warmed-up potatoes and salt herring 'and small beer with sops and syrup in an earthenware dish, which stood bubbling on the oven', or sometimes only a bit of bread to 'wash down with small beer'. An old fisherman's wife from Baskemölla told me quite unprompted in 1958, that this dish was known as *drickabrö* small beer bread'. Both the word and the dish have their counterpart in Denmark, where *øllebrød* (literally 'beer bread', a mixture of rye-bread, sugar, and small beer) is still drunk every morning on some farms in Fyn and Langeland. The Danish historian Troels-Lund has correctly described this dish as a precursor of today's *smørrebrød* (literally 'butter bread'). It was a common dish in the mid-sixteenth century, mentioned in the navy's food regulations, and was almost certainly known in Skåne at the same time.

A cross between the cold and hot mixtures of small beer and milk, was known as *ölost*, literally 'beer cheese', a mixture of hot small beer and hot milk. The addition of the small beer curdled the milk. This mixture occurred in Skåne but was much more common in Småland. Unlike all the other mixtures we have considered, *ölost* was something of a festive dish. It was always served when people gathered in a house of mourning early on the morning of a funeral. In Småland an '*Ölost* morning' was synonymous with a funeral.

THE DIETARY CONTEXT

The dominance of small beer in the diet in Skåne is connected with the cultivation of barley. Outsiders visiting the plains of Skåne said that the cornstacks stood so close together that it was scarcely possible to pass between them. Yet the Scanian barley did not keep well, and therefore lent itself better to export in the form of malt. In the towns every large house had its malting house, and in the plains the peasant farmers had bake-houses which contained not only an oven but also a malt kiln. People brewed as often as they baked. Yet it was only for major festivals that they brewed beer. The rest of the time they made small beer, which usually fermented within a week.

The plentiful supply of small beer in the countryside went hand in hand with a grave shortage of milk. This was due to the scarcity of pasture for cattle, since almost all cultivable soil was used for growing grain. The scarcity of milk is seen in the fact that it was small beer and not milk, that was drunk along with porridge in Skåne in former times. Reminiscent of this is the custom whereby old people drink small beer, or small beer mixture instead of milk, with the Christmas porridge (rice pudding).

The dominant role of small beer in the diet is also evident from the way it was used in its hot or cold mixtures as *brölura*, literally 'bread-lure', a way of making it easier to eat bread when there were no other accompaniments. Sops are nothing unique for Denmark and the former Danish provinces in Sweden. In former times the combination of bread and broth (*sod*) was a common everyday Swedish dish, but now it is served only at Christmas in the form of *dopp i grytan*, where bread is dipped in the stock from the ham (rather like the old English 'sop in the pan'). Comparable traditions were found in regions with a plentiful supply of milk; we may mention here the *blöta* in Norrland, consisting of crumbled flat bread soaked in milk. Wine had a comparable function in wine-growing countries, and we should not forget that aquavit was used in this way in Sweden in the eighteenth century and for some time in the nineteenth. Instead of beer people used aquavit to wash down dry bread at two or more meals each day. Eva Wigström even mentions 'aquavit soups', a direct equivalent of the 'soups' made with sops soaked in small beer or milk.

How are we to interpret the occurrence of the mixture of small beer and milk in Skåne? Mixing the two beverages was undoubtedly a necessity to save milk. If this was the case in the days of peasant self-sufficiency, then it became even more urgent when the monetary economy came. Several examples of prices show that small beer was much cheaper than milk. In Lund, around 1910, the price per litre for sweet milk was ten öre, for skimmed milk five öre, and for small beer only two öre. In northwest Skåne the price some years later was 15 öre per litre for milk, and seven to eight öre per litre for small beer. The difference in price must have helped to make small beer mixture the staple drink of the urban working class.

The old system of household management meant that nothing was to be wasted, not even the last drop of milk or small beer. Many informants even see here an explanation for the mixing of milk and small beer. The small beer had to be brewed or bought in fairly large quantities, and it easily went sour, especially in the summer when the alcohol started to ferment and became vinegar. Mixing in milk gave it a milder taste, and nothing went to waste.

That the admixture of milk was intended to improve the taste, is also recorded from Denmark, where small beer mixture did not otherwise occur. Yet this explanation is not entirely satisfactory, when we consider that it was an everyday drink with meals in Skåne and southern Halland, where there is nothing to suggest that milk was only mixed in when there was no fresh small beer. On the contrary, we are told that the mixture was never as tasty as when it was made from newly-brewed small beer and newly-strained milk.

The informants' responses maintain that the small beer gave the milk a fresher, slightly sour taste. This may be true when the mixture used sweet milk. However, in earlier times when the mixture could also be made from sour milk, the role of the small beer must have been the reverse. It is thus misleading to attach too much importance to modern-day attitudes to taste in seeking an understanding of the origin of the mixture. Nevertheless, it appears as if considerations of both taste and economy were satisfied by the ability of the mixture to refresh without increasing the thirst, the way unmixed small beer does.

Both the hot and the cold mixture have their place in a dietary pattern. Small beer sup is a variation on the theme of 'bread soaked in liquid'. Along with the unmixed liquids like broth, milk, small beer, or aquavit, there were also hybrid forms, of which small beer mixture is an example. These mixtures must be seen in their ecological context.

Beer, small beer, and aquavit have a natural connection with grain-growing areas, while milk is characteristic of regions with extensive pasture lands. In Germany, therefore, beer was the basis of a number of dishes, like *Biersuppe* and *Warmbier*. A linguistic relic of the importance of beer is seen in the German word *Kindelbier*—a christening feast (synonymous with Swedish *barnsöl* and Danish *barsel*, originally the ale brewed to celebrate the birth of a child)—which is

148 Milk and milk products

found only in northern Germany, an area with long traditions of brewing.

The ecological context can also account for the occurrence of certain mixed beverages. This is true, for instance, of the drink that was common in much of Sweden, especially in Norrland and Dalarna, known simply as *blanda* 'mixture'. It consisted of milk mixed with water, and it quenched the thirst of mowers working in fields far away from home. It was particularly suitable in that it made the transport of liquid for the workmen easier since only half of the required liquid needed to be transported to the fields, where water was already available.

There are other mixtures which cannot be understood outside their cultural and ecological context. One such is the hot aquavit soup mentioned by Nicolovius, in his description of life in Skytts Hundred in Skåne in the nineteenth century. It was made of aquavit, wheaten rusks, sugar, and water, and it was served to women at church festivities. Ignoring the sugar, which was a typical ingredient of alcoholic drinks for women, we see that this is a finer imitation of small beer sup. This is not to imply that the imitation originated in Skåne. There are several German equivalents, and it is quite possible that aquavit soup is a loan from there. North Germany also had its beer soups, and the German ethnologist Günter Wiegelmann has shown that the grog of north-western Germany is modelled on these older *Biersuppen* and *Warmbieren*.

As a drink with meals, beer and aquavit were replaced in the nineteenth century by coffee. As Wiegelmann has demonstrated, however, this change was not uniform, and it was not always the case that the older drink gave way to the newer one. In the north of Germany coffee became an everyday drink, enjoyed in the morning, for example, while people in the south continued to eat *Suppe* in the morning, and had coffee only on special occasions. In an intermediate area there was a hybrid form—*Kaffeesuppe*. The novelty, coffee, was introduced into the dietary pattern without replacing the older soup.

A close equivalent of this mixture is the Swedish *kaffegök*, literally 'coffee cuckoo', coffee laced with aquavit. It has been suggested this was spread by seamen from the west coast. In Skåne it appears to have been a common and popular drink in the 1840s. Eva Wigström describes how the master of the house in Rønneberg Hundred transformed his morning coffee into a 'cuckoo' by pouring in a glass of aquavit. In my opinion, this custom should be seen against the background of the peasant's earlier practice of starting the day with a dram of liquor and a bit of bread. The new custom did not mean that the old one was abandoned; instead there was a gradual transition. Later the 'coffee cuckoo' became a drink offered to guests, and it attained great popularity in Skåne and southern Halland, perhaps because people there were used to such mixtures.

The natural interpretation after these comparisons is that small beer mixture has its origin in the encounter of two cultures, the older milk-producing animal husbandry, and the younger grain- and malt-producing tillage. This may justify seeking the origin of the mixture as far back as the late Middle Ages. Compared with Denmark, where they knew *øllebrød*, but not small beer mixture, Skåne appears to be an archaic area, retaining the mixed forms.

It cannot be ruled out, of course, that small beer mixture arose later as a result of a dwindling supply of milk, perhaps caused by the waves of *rinderpest* that ravaged Skåne. It should be remembered, however, that Barchaeus, as we have seen, described the mixture as a general evening dish in Skåne as far back as 1773. From his account, and other early descriptions, it appears that small beer mixture was originally made with sour milk, not sweet milk, as in more recent times. This feature also suggests that the custom is very old.

THE REGRESSION

The informants all relate that both small beer mixture, and small beer sup, have become much less important in the diet than they once were. In particular, breakfast in the latter half of the nineteenth century ceased to consist of small beer sup, which was replaced by coffee with bread and butter. From having been a staple dish, it became one of several possible alternatives for dessert, and also a way to eat up the remaining dry crusts of bread. Small beer mixture, which was also originally eaten with a spoon, became a refreshing drink, but in this function it was later replaced by coffee and modern soft drinks.

The economic changes in the second half of the nineteenth century brought about a boom in milk production, partly aided by the introduction of the dairies. There was no longer a shortage of milk on the plains. At the same time, the abandonment of self-sufficiency meant that people stopped making their own small beer, which they instead bought from the breweries. This led to a reduction in the difference in the price of milk and small beer, so that it was no longer worthwhile mixing the two to save money. Increased prosperity also meant that people could afford to throw away stale bread and sour small beer. With more fresh food to eat, it was no longer necessary to have such a refreshing drink to accompany meals.

Along with all this there came a change in tastes. In the days when small beer mixture and small beer sup were eaten from the same dish, everyone had to make do with the food that was served. When this practice ceased, and people started mixing milk and small beer in the proportions they liked, it opened the door for a completely new individualism as regards taste. The new alternatives were a precondition for the change in tastes.

The varying attitudes to small beer mixture which are reflected in the informants' responses, clearly show how this dish is in a phase of disintegration. Those who express a negative attitude, not to say disgust, have only ever had occasional contact with the drink, and can even be put off by its colour. The positive evaluations come from informants who continue to enjoy the mixture, or from those who identify it with their childhood and birthplace, and therefore describe it in nostalgic terms. It is only in exceptional cases that people who have moved to other parts of Sweden have taken the custom with them.

These changes have combined to make small beer sup virtually disappear from the diet, while small beer mixture survives only as 'Grandfather's drink' (one name under which small beer is sold today), drunk only by the older generation, or only on special occasions. The mixtures have disappeared even more completely from the cookery books. It may even be questionable whether the essential component of both these mixtures—small beer—is likely to survive. In many households, especially among younger people, small beer is drunk only at Christmas, a typical sign of a dietary custom on its way to becoming a dietary relic.

In principle, I believe that food and drink mixtures are of the greatest interest in the study of food culture, since they give us the opportunity to investigate how different cultural elements meet in time, place, and in socio-economic contexts. I hope that I have also shown that cultural encounters like these have repercussions on people's tastes and preferences. We need more than just descriptions of the distribution of different food elements; we also need analyses of the cultural mechanisms behind them. Only through analyses of this kind can we learn something new about people as cultural beings in our study of food and foodways.

NOTE

A Swedish version of this paper, with full references, has been published in *Gastronomisk kalender* 1972.

1. Translator's note: *sowl* is an obsolete and dialectal English word used here to translate its Scanian dialectal cognate *sul* (Standard Swedish *sovel*). In both languages the term means 'any kind of food eaten with bread, as meat, cheese, etc.' (*Oxford English Dictionary*). Translation by Alan Crozier.

14 Porridge consumption in the Netherlands: changes in function and significance

Jozien Jobse-van Putten

This paper, dealing with the different functions and changes of position of porridge in Dutch eating habits, represents one aspect of a larger, ongoing research project on meal patterns in the Netherlands.[1] For the period 1900–1925 it is based on archival sources, i.e. the responses to two different ethnological questionnaires on food issued by the Ethnological Department, P.J. Meertens-Instituut, Amsterdam, in 1976 and 1985.[2] Historical literature on nutrition indicates the significance and role of porridge in the Dutch diet prior to 1900. Both types of sources place the modern developments in the consumption of porridge in the Netherlands in context. The aim of this paper is not only to indicate a number of patterns in the development of porridge as a common dish in the Netherlands, but, more importantly, to discover whether, from this sketch of the life-history of a single dish, determinants of nourishment behaviour can be discerned that are valuable for nourishment research in general.

THE CONSUMPTION OF PORRIDGE, 1900–1925

From the questionnaire replies it seems that in the first quarter of the twentieth century, porridge was a dish of great importance in the daily meal system of the Dutch countryside. That this should be so is not surprising. Porridge is a partly solid, partly fluid, cooked dish that is made from a variety of grains (sometimes even from grain products such as bread and rusk), with a fluid ingredient such as milk, buttermilk, beer, water, or a combination of these liquids. All these ingredients have always been available in large quantities, and in a number of varieties, in the Netherlands. Furthermore, porridge is an inexpensive and simple dish that makes no great demands on cooking equipment and facilities, culinary skills, or fuel quantities. Thus, from a historical standpoint, the important role of porridge in the Dutch diet is understandable. But the Netherlands has not been unique in this connection, since, in numerous other countries in Europe, porridge has also been a popular dish in the past.[3] No doubt its popularity everywhere owes something to the simplicity of its preparation since, as stated, it can be made even under relatively primitive conditions and with simple utensils.

Evidence of the importance of porridge in the Dutch diet at the beginning of this century is the frequency with which it was eaten. In rural areas of the Netherlands at that time it was commonly eaten once a day, but twice a day was not exceptional. Furthermore, there were also numerous varieties of porridge which could differ not only in ingredients, but also in consistency.

But, perhaps, the important role that porridge played in the diet of the Netherlands in the period 1900–1925 is evident above all from the variety of ways in which the dish could be, and indeed was integrated into the different meals. Porridge could feature in every meal of the day; it could be used as a main dish, or as an appetizer, as a side-dish, or as a dessert, and it could also be combined with various other foods, or dishes, such as bread, potatoes or pancakes. Thus, at the beginning of this century, porridge was a diversely served, multi-functional dish, that existed in numerous variations.

Although porridge featured in the daily menus nearly everywhere in the Netherlands, nevertheless, there was substantial variation in the distribution of porridge dishes between the different meals. These differences were partly social and partly regional in nature. Thus, among the peasantry in one region, porridge was served as a regular extra meal before going to bed, while the townspeople of the same region had such an extra meal only infrequently. In another region, porridge functioned primarily as the second course of a hot meal among the peasantry, while elsewhere, and also among other social groups, porridge could be combined with bread to form a meal. Since two hot meals were often consumed daily in rural areas during the period 1900–1925, the use of porridge could vary between these two meals. Thus, in one region, porridge could be used exclusively during the hot midday meal, while elsewhere it could be consumed at midday, and again in the evenings when hot meals were eaten.

THE CONSUMPTION OF PORRIDGE IN PREVIOUS CENTURIES

In the historical literature of nutrition, porridge is considered a dish that has been consumed in the Netherlands from at least the Middle Ages, especially among lower social groups, and in the countryside.[4] Accordingly, in the general characterisations of Dutch nutrition that exist for the past (and, unfortunately, these are only 'general' characterisations), porridge is not overlooked.[5] However, we are but poorly informed about the social, regional, and functional characteristics of the dish in previous periods, and these aspects of porridge consumption must, therefore, be reconstructed.

Menu lists from banquets, charitable, and other institutions, from the past, offer a possible means of measuring the place of porridge in the diet in earlier centuries.[6] The analysis of this kind of source material poses certain problems of interpretation, however. One is relying on the names of dishes given in the menus, but these names, just as the dishes themselves, were sometimes influenced by regional or social dialect and custom, and they may even have been completely lost in the course of time. Moreover, the spelling can complicate the interpretation;[7] this means that we often have to guess the type of dish behind a certain name. Even popular dishes such as pearl barley, and groats or grits,[8] are difficult to classify. That pearl barley and groats were often made into porridge or gruel, is confirmed by terms such as Dutch *gortpap* and *gruttenbrij* (barley gruel and groats porridge, respectively). Yet this does not mean that we can interpret every mention of pearl barley or groats as a porridge dish.[9]

On the basis of the menu lists, and in combination with available literature on nutrition, it can be asserted that porridge was also eaten in well-to-do circles, although less frequently than among poorer social groups. Furthermore, it seems that porridge was not often served at banquets, since it was never considered a prestigious dish. Whenever porridge was eaten in better circles, it was mainly a luxury sort made from expensive grain or grain products (wheat, rice, or whitebread), using cream or whole milk (instead of buttermilk or water), and sometimes enriched with eggs, or even almonds.[10]

For poorer groups, however, porridge was much more important. To know more about their food I have used menu lists from charitable and other institutions. It can be cautiously assumed that the food in such institutions showed similarities to the fare of broader sections of the population.[11] Throughout the centuries, porridge graces the menus of hospices, orphanages, and houses of correction, mostly on a daily basis; and in student eating halls, also, porridge was far from unknown. It is striking that porridge could be served in the morning, at noon, and even for evening meals. The role porridge played within the meal is also remarkable; in many cases dishes were served together, irrespective of whether they were sweet or savoury,[12] that is to say, porridge could be served and eaten together with meat, as a side dish. But it could function also as the main dish, especially in the evening meal, but also in the morning meal. It seems that porridge functioned far less often in the past as a dessert.

DETERMINANTS OF PORRIDGE CONSUMPTION

In the previous sections we have demonstrated the considerable variations in the kinds and uses of porridge around the turn of this century. The frequency with which porridge was eaten per day, the meals in which it appeared, the function the porridge had within the meal, the sort of porridge that was consumed, and the dishes porridge was combined with, are all matters closely related to economic and social milieu, to region of residence, group preferences, personal taste, the day of the week or the year (a weekday, Sunday, or holiday), and even to a combination of these factors. It was also important if the food itself (wholly or partly) was produced by the consumer. The significance of some of these factors will now be briefly expanded upon for the period 1900–1925.

Those who were in the habit of eating porridge twice a day did not necessarily eat the same sort of porridge on both occasions. There were numerous possibilities for variation, but not everyone could afford them. The choice that one made was often determined, in part, by living conditions, such as wealth and access to ingredients.

Those who had their own land and livestock generally ate porridge made from the grain and milk products that they themselves produced. A dish such as porridge had extra significance within situations of self-sufficiency, however, for it was a suitable means of using up left-over products such as buttermilk and skimmed milk. Despite the undoubtedly close relationship between an individual's daily and seasonal food and his food production, culinary freedom was, of course, possible in the preparation of the food, and this was expressed especially in the choice of dishes. From grain one could also prepare, not just porridge, but also bread, pancakes, and other flour products (such as pudding). The frequency with which these foods were eaten, and the quantities in which they were consumed, differed among households, regions, and also time periods, all of which could be determined by diverse factors.

One of these factors was the method of preparation. Porridge was cooked, and cooking (or boiling) is the simplest means to make grain digestible. The porridge of the past might often curdle or form lumps, but these shortcomings received little attention at that time. More important was the fact that, as we have seen, the method of preparation required less expertise, equipment, fuel and time, than was the case with other grain dishes, such as bread, for example. Therefore, the task of baking bread had been passed on over the ages to a craftsman, the baker. This process of specialisation was completed already around the turn of the century in the cities and in some rural areas, though elsewhere it came to an end only gradually in the course of the twentieth century. For those who ceased to bake their own bread, bread—in comparison with other grain dishes—became an expensive food. Therefore rules existed in some families which regulated the use of bread. Thus bread was sometimes served in appropriately measured quantities, while porridge could be eaten before, together with, or after the bread, in unlimited quantities.

Porridge was also eaten by the rural population who had little or no chance of self-sufficiency. The ingredients for porridge were easily obtained in rural areas, and, in general, they were relatively inexpensive.

Since the selection of ingredients did not depend on a self-sufficiency situation, personal preference could play a larger role in the acquisition of the grain and milk products needed for porridge. Nevertheless, the type of porridge eaten by this group was also dependent on socio-economic conditions. In choosing from the many different sorts of ingredients available they would have been influenced, above all, by considerations of price.

Yet, more than just socio-economic factors played a role. Porridge was a dish much liked by the rural population, at least on work days. The pervasiveness and popularity of porridge in the past is demonstrated by the many proverbial sayings in Dutch referring to porridge. However, in the period 1900–1925, the strong position of porridge in the diet began to weaken somewhat. On Sundays and holidays, when no work needed to be done, the choice fell often on Dutch 'pudding' (in the Netherlands 'pudding' is a very stiff kind of custard made in a mould which can be turned out on a serving plate), a fruit purée, or a somewhat richer sort of porridge, such as rice milk or rice pudding. There were still groups that ate traditional porridge also on Sundays and holidays, but many tried to eat something more festive on these special days. Thus on these days, sugar or currants were often added to the rice milk, Dutch 'pudding', or fruit purée. The weekday porridge virtually always lacked this type of refinement, and in many cases, nothing apart from salt was used to enhance the flavour. The use of sugar, especially brown sugar, was on the increase, but hardly prevalent. More common was the use of treacle, especially in (sour) buttermilk porridge, but this addition was also not necessarily the rule. Fat was also sometimes added to the porridge in the form of butter or dripping. Whether one actually permitted oneself a somewhat richer meal on Sundays and holidays depended of course on one's economic situation, as well as the importance placed on luxury. It will be clear, however, (but this observation is not new[13]) that the motivation to eat more tasty, luxurious, or prestigious things on Sundays and/or holidays, can be considered one of the compelling factors for development and change in food habits.

The reason why porridge was, in fact, a weekday favourite lay, above all, in the significance which the population attached to the dish. It was believed that porridge was a dish that enabled one to work well. The inhabitants of the countryside saw this relationship as a direct one; the distinction between 'work' and 'life' was often not very great. If one wanted to be able to live, one had to work, and if one wanted to be able to work, one had to eat. The need for energy was great, since work then required greater physical strength than now. For these reasons, large quantities of heavy food were appreciated. Substantial fare—that's what people wanted! Porridge was a dish that fitted the bill as most porridges were thick and hearty, and sometimes they were made so thick that a spoon could stand in them! Occasionally even beans or peas were added during the cooking process to thicken them. For the same reason, extra bread was crumbled into the porridge bowl. It is clear that refinement was not the aim; coarse and heavy food was considered delicious. 'Filling the belly', as it was called, was the most important factor. Even when porridge was eaten after a meal, it rarely had the significance of a modern dessert. Porridge hardly ever functioned in the past as a small, light, or sweet end to a meal; it was mainly a heavy and hearty dish. Since porridge was consumed, moreover, in large quantities, it constituted a fully-fledged second course that, as part of a meal, could scarcely have been left out; without porridge one had not eaten enough and the meal was incomplete.

So, culinary delights and considerations of taste did not mean much to the majority of the rural population around the beginning of this century; rather 'a lot of food', or simply, quantity, featured considerably higher on their list of priorities. However, interest in more tasty foods did increase somewhat around this time, so, as we have already seen, on Sundays, a 'lighter' fare, as it was called, gradually began to make its way onto the dining table. But the rather condescending language used to refer to the new fare reflected an ambivalent attitude to it, an attitude typical in such new situations.

Porridge was not only greatly valued because of the belief that it fortified one for a good day's work, but also because it was considered a healthy food, not so much because it was a dish prepared with good ingredients (grain and milk products), but rather because substantial foods were recognised as healthy foods. In contrast to what we are now accustomed to, in earlier times people urged each other to eat heartily. Here, too, simple causal connections were made; you must eat well (i.e. a lot), otherwise you will get sick. Those who were fat or stocky, were thought to look healthy, while those who were slim were called 'gaunt'. People thought the latter looked 'bad' and were susceptible to all kinds of illnesses. People then were still very apprehensive of disease; and rightly so, for contagious diseases such as tuberculosis, polio, and diphtheria still broke out regularly as a result of relatively primitive living and working conditions, poverty, poor to mediocre health care, lack of hygiene and knowledge thereof.

Eating a lot of heavy foods was highly thought of among earlier rural populations for more than one single reason. It was also a way to express one's resistance to the poverty that was then directly visible to all; by eating a lot one could distance oneself from poorer groups. This attitude can be understood as a consequence of the great poverty of the nineteenth century from which some had only just escaped, and from which others were still in the process of withdrawing. That the memories of poverty were still alive, can be inferred from the way people continued to think about eating for the following decades. One did not eat just to be satiated; no, after a meal one wanted to feel the walls of the stomach stretch! Porridge was a suitable dish for this as 'It filled the last cracks', so to speak. Hunger was associated with poverty, and no one wanted to suffer from hunger, or to be considered poor. These who could eat plentifully and substantially indicated that they were not poor, and thus the desire and ability to eat a lot was evidence of the progress that had been achieved in a human life.

MODERN DEVELOPMENTS IN THE CONSUMPTION OF PORRIDGE

Today, in the Netherlands, porridge is rarely cooked at home, and the skills involved in cooking the various varieties are largely unknown. A number of grain products that were used for porridge (such as pearl barley and buckwheat groats) have not been available for a while, but because of recent interest in alternative foods and the health movement, some are once again available for purchase; yet they are seldom used nowadays for the purposes of making porridge.

The preparation of porridge has been taken over (in various ways and in various phases) by industry. Initially, factories imitated the homemade porridge. In some regions, the local dairy-factory produced pearl barley porridge, flour porridge, etc., which could be purchased loose from a farmer's cart, while elsewhere they were available in bottles.

Industrial production also brought on the market quick-cooking grain products (in the form of cream of wheat), such as oatmeal, for those who still wished to make their own porridge. In addition, from the beginning of this century, the food industry offered numerous 'pudding'[14] and custard mixes of varying flavours. In the 1950s these mixes were perfected to the point where the milk no longer needed to be boiled;[15] and one could purchase custard in bottles also.

Custard and Dutch 'puddings' are foods which originally required certain skills and expertise in their preparation which were not to be found in most Dutch homes of a century ago. They were thus eaten primarily by the upper class, and the urban middle class who had the services of maids. These desserts were also purchased by some craftsmen. The instant products which the food industry brought on to the market, quickly captured a place in the Sunday meal of the wider urban and rural populations. These prepared products functioned almost exclusively as special desserts. Thus it appears that industrialisation (together with the rising standard of living) contributed to the democratisation of nourishment.

Changes have also taken place in the use and the function of porridge. While porridge could be served in the past as an appetiser, a side dish, a main course, or a dessert, nowadays it is known primarily as a dessert. However, its function as a main course has not been totally lost; some people today still eat porridge for breakfast. Aside from yoghurt, ready-made products such as cornflakes, muesli, crispies, and weetabix are used.

Dessert, especially, has another meaning today than it used to have. It is no longer a large, filling course, without which one feels that one has not adequately eaten. Today, a dessert is considered a small, sweet end to the meal, and as far as porridge is concerned, it now functions only as a 'sweet'; it is something small and extra in a meal that, in theory, could do without it. And in contrast to the beginning of the century, 'sweet' flavours predominate. The close alliance between sweetness and desserts was common in festive and banquet dishes of earlier centuries, yet salty dishes, such as pasties, were eaten then after the main courses as well.

Developments in desserts have increased enormously in recent years. Traditional pearl barley and flour porridge are also no longer available. They were overwhelmed by countless varieties of yoghurt, curd and other milk products—all of which have been refined and enriched in endlessly different ways, with eggs, whipped cream, nuts, currants, raisins, and other fruits, as well as with flavours, colours and thickeners. It is striking, however, that in these dairy desserts, the grain element—an integral element in earlier porridges—is no longer represented, or at least not in any significant way. Earlier porridges were combinations of grain and milk ingredients, and of the two, it was grain and not milk which was actually the most characteristic ingredient, since beer or water could also be used instead of milk. In the modern dessert it is precisely the grain element that is not, or is just barely, to be found; instead of a combination of grain and milk products, it is now in fact only a milk product.

As we have shown, the function of porridge has narrowed almost exclusively to that of a dessert. Only in its (subordinate) role as a main dish at breakfast, is the earlier composition of porridge still recognisable. For this function, a high nutritional value is required. Since grain forms the basis of products such as cornflakes, muesli, crunchy, etc., these 'porridges' can function as complete morning meals. In general, therefore, one can say that in recent years in the Netherlands porridge has developed into a completely new product and that traditional porridge has become a dish of the past.

CONCLUSION

A number of the recent developments in the consumption of porridge in the Netherlands, which apply in more general ways to changes in general nourishment, are not really specific to the Netherlands, since they occur also in a number of western countries. Of relevance here, for example, are the enormous developments in industrial food production. Because self-sufficiency has faded into the past, and multinationals such as Unilever, Danone, etc., produce and distribute their products worldwide, the character of not only regional, but also of national foods, has become less significant. This trend toward internationalisation is expressed in the supply, as well as in the language of food, for nowadays the names of numerous products appearing in the Dutch market, such as cornflakes or muesli, are not translated into Dutch anymore.

Since the laws of supply and demand apply in industry, the supply as well as markéting reflect the demands of society. Thus the need for ease and efficiency, which has fully penetrated society in the course of this century, has led to the use of all sorts of instant and ready-made desserts instead of home-cooked porridge, which is no longer necessary. In the whole area of nutrition, this trend can be seen in the growing importance of wholly, or partly prepared products, the so-called convenience foods.

There are other trends occurring in porridge consumption that can be observed more generally in the development of food products. Desserts have grown into products characterised by a great degree of refinement. Because modern employment and contemporary social conditions are physically less demanding, the nutritional value of what is eaten does not play such a pre-eminent role. Instead of the thicker porridge of yesterday, light and sweet desserts, enriched in all sorts of ways, are served today. The flavour, appearance, and decoration of foods, and the enjoyment to be derived from them, have become the important factors.

Another factor in relation to attitudes to food is the higher standard of living brought about by the welfare state. In all aspects of life, there are greater demands, and in relation to food, the demands cannot be met anymore in quantitative ways, but rather in qualitative ways. Just as in the past, group differences still occur, but, nowadays, these are not only differences between social groups, but differences in age and in life-style within social groups.

Present-day thinking is also influenced by the shift from quantitative to qualitative attitudes to food. Thus, we shudder at the thought of a spoon standing in thick porridge, as it was supposed to do a half century ago. Porridge no longer has to 'fill', and it is even better if it does not. The health movement has created a demand for lower calorie food, and moreover, slenderness has become the norm. The following may be regarded as a general rule today: the lighter, the better! (though objectively seen, many modern desserts do not meet this requirement).

Furthermore, nowadays, there is great willingness to accept cultural enrichment, and the increasing pluralism of modern society has led also to a different attitude towards food; adventurism has become an aspect of nourishment behaviour. In the past, people ate especially what they were familiar with (neophobia), and although this attitude has not been totally lost, nowadays there is an eagerness for new foods, foreign ingredients, different ways of cooking, new combinations of tastes etc. (neophilia). Love of variety (which is supported by the health movement), and of what is new[16] characterise present-day attitudes to food.

In conclusion, we wish to assert that a specific dish, such as porridge, shows that food has a certain expressiveness in the society within which it functions, and that taste and group preferences change under the influence of altering social conditions. In times of poverty, when diseases were often threatening, and great demands were made on the human body, thick porridge was a favourite dish that was counted on to fortify one for a hard day's work, and to keep the body healthy. One tried to distinguish oneself from the less privileged class, above all, in quantitative ways. But today, when work demands less physical strength, and when there is what has been called a rise in the diseases of 'prosperity', such as heart disease, the number of nourishing ingredients has to be limited. Since being overweight has acquired certain adverse social and even stigmatising connotations,[17] people today can only distinguish themselves from others in qualitative ways,[18] that is, by seeking refuge in more expensive, more tasty, and more prestigious foods. The traditional porridge dish which has become a contemporary modified dessert food is one of the ways to realise that wish.

NOTES

1 This project is being conducted by the Ethnological Department, P.J. Meertens Instituut, Amsterdam.
2 Questionnaire 45 (1976) and 55 (1985). For the precise questions of these questionnaires dealing with porridge, see, *De volkskundevragenlijsten, 1–58 (1934–1988) van het P.J. Meertens-Instituut*. Ingeleid door A. J. Dekker; voorzien van een register door J. J. Schell, Amsterdam 1989 (Publikaties van het P. J. Meertens-Instituut 12), 220–221, 266.
3 See, for example H.-J. Teuteberg and G. Wiegelmann, *Der Wandel der Nahrungsgewohnheiten unter dem Einflusz der Industrialisierung*, Göttingen, 1972, **33**, 135; I. Arnö-Berg, 'Från kvällsgröt till grötfrukost', *Mat*, [Special issue of] *Fataburen* 1989, 65–73, 173–174.
4 F. E. J. M. Baudet, *De maaltijd en de keuken in de Middeleeuwen* Leiden, 1904, 101. L. Burema, *De voeding in Nederland van de Middeleeuwen tot de twintigste eeuw*, Assen, 1953, 36, 59, 72; A. van't Veer, *Oud-hollands kookboek*, Utrecht/Antwerpen, 1966, 12, 25; R. J. Bron, en I. van Houten, 'Interview met Professor Van Winter', *Voedselgeschiedenis*. [Special issue of] *Groniek* 95 (1986), 95.
5 See, for example, J. Scheltema, 'Geschiedenis van de dagelijksche kost in de burger-huishoudingen', *Geschied- en letterkundig mengelwerk*, vierde deel, II, Utrecht, 1830, 284; L. Burema, *op. cit.* 1953; H. van Nierop, 'Het tijdperk van het brood. Van 1500 tot omstreeks 1800', *Brood, aardappels en patat. Eeuwen eten in Amsterdam*, Eds. R. Kistemaker en C. van Lakerveld, Purmerend, 1983, 4–25; L. Boiten, *Eten om te leven, leven om te eten. Groningers aan tafel sinds de Meddeleeuwen*. Met medewerking van R. Nip en M. C. van der Sman, Groningen, 1986, 9, 13, 98.
6 I am basing my observations on only the limited number of menus that are cited in the literature; I have, accordingly, not aimed for comprehensiveness.
7 Thus, one comes across words like 'bierenbroot' and 'melkentweebak' (bread-porridge made with beer, and rusk-porridge made with milk, respectively), but sometimes also 'bier en broot' and 'melk en tweebak'. The latter can refer to porridge, yet also to two other dishes, namely bread (i.e. rusk) with which beer (i.e. milk) was served.

8 Pearl barley (Dutch: *gort*) and groats or grits (Dutch: *grutten*) were both grain products which could be distinguished by the way they had been processed, though they also could be from different grain sorts (for example, barley, oats, buckwheat). Although the designation was not identical in all regions, the difference can be generally assigned as: pearl barley (*gort*) is hulled grain, and groats or grits (*grutten*) is crushed grain.
9 Pearl barley and groats can be cooked to a mash and eaten in a liquid state as gruel or porridge; yet they can also be kept whole during the cooking process and then eaten in this form as an independent dish.
10 R. J. Bron en I. Van Houten, *op. cit.*, 95; C. G. Santing, 'Het toetje door de eeuwen heen', *Voedselgeschiedenis*. [Special issue of] *Groniek* **85** (1986), 142.
11 See, for example, H.-J. Teuteberg and G. Wiegelmann, *op. cit.*, 148–51; J. van der Maas en L. Noordegraaf, 'Smakelijk eten. Aardappelconsumptie in Holland in de achttiende eeuw en het begin van de negentiende eeuw', *Tijdschrift voor sociale geschiedenis* **9** (1983), 188–220. That food in closed communities could also differ from the everyday bourgeoisie fare is also shown by A.M. van der Woude. 'De consumptie van graan, vlees en boter in Holland op het einde van de achttiende eeuw', *A.A.G. Bijdragen* **9** (1963), 128, 136.
12 A. van't Veer. *Op. cit.*, 1966, 12.
13 See for example, G. Wiegelmann, *Alltags- und Festspeisen. Wandel und gegenwärtige Stellung*, Marburg, 1967.
14 'Pudding' as already stated, is, in the Netherlands, a very stiff sort of custard, which can be turned out of a mould on a serving plate.
15 Santing, *op. cit.*, 146–149.
16 M. Visser, *The rituals of dinner. The origins, evolution, eccentricities, and meaning of table manners*, New York, 1991, 42–44.
17 A. H. van Otterloo en J. van Ogtrop, *Het regime van veel, vet en zoet. Praten met moeders over voeding en gezondheid*, Amsterdam, 1989, 21–46.
18 For the shift from the quantitative to the qualitative as a means of social distinction in the nourishment of Western Europe, see especially the study of S. Mennell, *All manners of food. Eating and taste in England and France from the Middle Ages to the present*, Oxford/New York, 1985.

SELECT BIBLIOGRAPHY

A. Bos, *Handboekje bij de beoefening van het zuivel-bedrijf in Zuid-Holland*, Den Haag, 1909.

H. B. Hylkema, *Historische schets der zuivelbereiding*, Leeuwarden, 1912; 3rd. printing.

V.R. I J. Croesen, *De geschiedenis van de ontwikkeling van de Nederlandsche zuivelbereiding in het laatste van de negentiende en het begin van de twintigste eeuw* (Dissertation, Landbouwhogeschool Wageningen), 's-Gravenhage 1931.

C. Schiere, *Het bedrijf van de zelfkazers in Utrecht en Zuidholland*, Utrecht, 1938.

M. J. L. Dols, J. Sevenster, *Productie en bestemming van melk in Nederland*, Rotterdam, 1950.

J. Sevenster, *Productie en bestemming van melk in Nederland II. Een confrontatie van twee studies*. (Dissertation, Landbouwhogeschool Wageningen), Wageningen 1953.

G. A. Kooy, *De zelfkazerij van Midden-Nederland: Een onderzoek naar haar voortbestaan*, Assen, 1956.

J. A. Geluk, *Zuivelroöperatie in Nederland; onstaan en ontwikkeling tot omstrecks 1930*, 's-Gravenhage 1967.

M. Wigbout, *Het gebruik van melk en melkprodukten volgens een voedingsonderzoek in Nederland bij tweeduizend huishoudingen in 1966 en 1967*. (Dissertation, landbouwhogeschool Wageningen, Parts I-II), Rotterdam, 1972.

162 Milk and milk products

Beeld van de Nederlamdse zuivel; een structuurstudie, Rÿswÿk 1974.

H. de Waard, *Het gebruik van melk en zuivelprodukten in de menselijke voeding; lezing gehouden tijdens het XXste Internationale Zuivelcongres*, (Parijs, 26–30 juni 1978), Rijswijk, 1979.

M. van der Marck, J. Slot, *De geschiedenis ener melkinrichting; een eeuw consumptie melk 1879–1979; jubileumuitgave*, Amsterdam, 1979.

J.G. Termorshuizen, *Het consumptiegedrag met betrekking tot melk.* (Dissertation, Landbouwhogeschool Wageningen), Wageningen, 1982.

M. S. C. Bakker, *Boterbereiding in de late negentiende eeuw*, Zutphen, 1991.

Stoomzuivelfabrick Freia en de onlwikkelingen in de Nederlandse zuivelindustrie, 1850–1870, Aenkem 1992.

Part IV
Ice-Cream

15 Ice-cream in Europe and North America

Laurie Jo Shapiro

When one enters a supermarket or an ice-cream parlour in the United States of America today, one can easily understand why ice-cream is considered an 'American Tradition.' The urge to create unusual and delightful ice-cream concoctions seems to be an inherent characteristic of the American ice-cream industry. The numerous variations take the form not only of 'ordinary' ice-cream, but of low-fat ice-cream, ice milk, sorbet, sherbet, yoghurt, non-fat yoghurt, ice-cream bars, water ice, popsicles, and many other frozen novelties. The varieties are also international, and include, for example, Italian *granita* and French *fromage glacé*, as well as regional, for instance *Philadelphia Ice Cream*. As for the flavours, the list is endless! But ice-cream itself has had a European ancestry, and this paper attempts to outline its history in Europe and in some North American cities, illuminating especially the Philadelphia ice-cream phenomenon.

ICE-CREAM IN EUROPE

There is little known of the early history of ice-cream. King Solomon (*Proverbs* 25:13) noted that drinks cooled by snow were enjoyed during the warm harvest season. The Arabs also used snow to chill *sarbat*, a mixture of fruit syrups and water. What the Italians call *granita*, a grainy water ice, is a variant of *sarbat* chilled for too long; it has since developed into *sorbetto*, a smoothly frozen treat. The Greeks and the Romans also enjoyed a cool treat, as they sometimes chilled their wine with lumps of snow. Nero Claudius Caesar (54–68 AD) was known to send slaves to the mountains to collect snow which was mixed with honey, juices, or fruits, to make flavourful ices. In the second century AD, a Roman food called *mecla* was consumed, and while there are no descriptions of this dish, it is assumed to have been a frozen milk product. The Venetian Marco Polo, on returning to his native Italy at the end of the thirteenth century, brought with him a recipe for a frozen dessert, closely resembling that of today's sherbet, which he had learned during his travels in Asia. Another culinary secret that Polo was thought to have revealed to Italy was the use of salt or saltpetre as a coolant.

Throughout the Middle Ages, ices and ice-creams remained delicacies that were consumed by the elite classes in various countries. The Florentine Catherine de Médici is said to have introduced these delicacies into France when she married the Duc d'Orléans who later became Henri II of France; she brought with her to France not only court attendants, but also the recipes for ices, ice-cream, and sherbet. In France, as in Italy, the elite classes endeavoured to keep this frozen delight to themselves. In England, too, in the early seventeenth century, ice-cream was a favourite dish of royalty; Charles I was said to enjoy ice-cream so much that he did not want to share the recipe even with his nobles.

In 1670 in Paris, the general public was able to enjoy ice-cream with the opening of Café Procope, on the rue de l'Ancienne Comédie. Francesco Procopio dei Coltelli, an Italian, was the first person ever to offer 'iced creams' and sherbet to the masses when he opened the first café or coffee house in Paris. Dei Coltelli, who was one of two hundred and fifty master *limonadière* who sold chilled lemon drinks, was inspired by Caffé Florian on the Piazza San Marco in Venice. Another Parisian café famous for its ice-cream was the Café Napolitain on the Boulevard des Italiens. The owner, also an Italian, created an ice-cream treat consisting of macaroon, rum, and cream, which he named after himself, Tortoni.

Elsewhere in Europe, as H.-J. Teuteberg has shown, the recovery subsequent on the Thirty Years War provided an economic and social climate conducive to the gradual adoption of luxury food items, for example coffee, tea, chocolate, etc., into the diet of most sections of society. Coffee houses—which often included ices and crèmes on their menus—developed as centres of bourgeois culture, and from around 1800, new kitchen and dining utensils came into vogue, and 'bourgeois cuisine' was adopted as the new social ideal. These structural changes can be clearly seen in the cookbooks of the period.

Already about the beginning of the eighteenth century cookbooks featuring ice-cream recipes were appearing in various European countries. *L'art de Faire des Glacés*, published in Paris around 1700, was an indispensable guide for preparing various ice-cream delicacies such as apricot, rose, caramel, and chocolate ice-creams. *L'Art de bien faire les Glaces d'Office*, (Paris, 1768) by M. Emy, also gave recipes for ice-creams. Early English ice-cream recipes can be found in *Mrs Eales' Receipts*, 'Confectioner to her late Majesty Queen Anne', (London, 1718) and Nathan Bailey's *Dictionarum domesticum* (London 1736). Another English cookbook, Mrs. Hanna Glasse's *The Art of Cookery Made Easy* (London, 1747) contained a recipe that explained the method and materials used to make ice-cream.

ICE-CREAM IN NORTH AMERICA: THE PHILADELPHIA CONNECTION

Just as the inventor of ice-cream is not known, the details surrounding the actual introduction of ice-cream into American cuisine is also unknown. The earliest reference in this connection is from 1744 when a dinner guest of Governor William Bladen of Maryland, wrote that the meal had included 'some fine Ice cream which, with Strawberries and Milk, eat most Deliciously'. Among prominent politicians who were fond of making, eating, and serving ice-cream were George Washington, whose household inventory included two pewter ice-cream pots purchased in Philadelphia; Thomas Jefferson, who created an early rendition of Baked Alaska (an ice-cream and cake creation that is covered with egg whites whipped to a meringue, and placed in an oven to brown); and James Madison's wife, Dolly, who served strawberry ice-cream at his second presidential inaugural banquet.

New York City was witness to some of the early developments in the history of ice-cream in America. In 1777, Philip Lenzi, a confectioner, placed an ice-cream advertisement in the *New York Gazette* on the 12th of May. This was the first time ice-cream had been advertised in a United States paper. In 1781, Joseph Corre became the first person to attribute to ice-cream its prominence in American tastes by placing an advertisement using capital letters, for his ice-cream product, in the *New York Post Boy*, and, in 1786, Joseph Crowe, whose family name is assumed to be an anglicised pseudonym for the Italian 'Corre', placed an advertisement, again in the *New York Post Boy*, for his ice-cream. These advertisements led two dairy historians to the following conclusions:

> Quite apart from the antiquarian interest which attaches to these advertisements is the fact that they were in a sense prophetic of the day when the adopted home of ice cream was to outstrip the countries of its origin in the degree it was to be developed and in popularity accorded it by Americans. (Dickson, 1972, 22)

Philadelphia, which in the 1700s, boasted numerous establishments selling ice-cream, was the hub for imaginative creators of amazing ice-cream concoctions and leaders in the ice-cream industry. Gray's Ferry and Gardens, a restaurant frequented by Governor Thomas Mifflin and by Mrs. Washington, had ice-cream listed on their bills in 1789. In 1798, Victor Collet placed a bilingual advertisement for his ice-cream shop:

ICE CREAMS
of all kinds
May be had at Mr. Collet's no. 127 North Front Street,
everyday between two in the afternoon and ten

in the evening. He also makes, when bespoke
all kinds of iced cheeses.
Des Glaces de toutes especes
Se trouvent chez le Sieur Collet, no. 127 North
Front Street, tous les jours depuis
deux heures apres midi jusqu'à dix heures
du soir. Il faut pour envoyer en ville
toutes sortes de fromage glacés.

In 1832, also in Philadelphia, Augustus Jackson, a black cook trained to make ice-cream at the White House, opened his ice-cream shop.

James Parkinson, who created a Philadelphia ice-cream dynasty, furthered the development of ice-cream, and its Philadelphia connections. An industrious restaurateur, Parkinson combined his tavern, Pennsylvania Arms, with his mother's confectionery shop, which allowed them to become quite successful in the culinary world. It may even be Parkinson's articles in the *Confectioners' Journal* (Philadelphia) that made 'Philadelphia Ice-Cream' a widely recognised confection.

In the course of the nineteenth century, ice-making machines and mechanical refrigerators were introduced which enabled further fundamental development of the ice-cream industry. Prior to the invention of mechanical refrigeration, ice-cream manufacturers had relied on the combination of ice and salt to cool their product, but now they could produce, store, and sell in bulk, resulting in growth in the ice-cream industry. Another development that benefited the ice-cream industry was Nancy Johnson's invention of the hand-cranked ice-cream freezer in 1846. But for some unknown reason she did not patent it, and two years later, William G. Young applied for a patent that for all practical purposes was Johnson's invention, and he marketed his machine as the 'Johnson-Patent Ice-Cream Freezer.'

The importance of ice-cream making equipment, moulds, tools, and machines did not escape the notice of Philadelphians. Valentine Clad, a Frenchman who immigrated to Philadelphia, established a family enterprise that became one of the major manufacturers of confectionery equipment in the United States during the second half of the nineteenth century. Clad was famous for his fine copper saucepans, elaborate ice-cream moulds, and professional confectioner's tools. As for ice-cream freezers, North Brothers Manufacturing Company and the Acme Can Company were just two Philadelphia manufacturers. The former company, the more well-known of the two businesses, produced many styles of freezers, including the 'American Twin,' 'Shepard's Jumbo Lightening,' and the 'Blizzard Freezer.'

Americans' love for ice-cream has been documented in various ways. *Godey's Lady's Book* (Philadelphia) in 1850, described ice-cream as one of the necessities of life. An 1851 cookbook from Philadelphia listed thirty-four different ice-cream recipes, along with eighteen recipes for water ices. Also in 1851, Jacob Fussel, a Baltimore milk dealer, set up the first wholesale ice-cream business in the United States, and as the ice-cream trade proved more successful than the milk business, he decided to concentrate on ice-cream alone. After Baltimore he invaded other markets with the exception of Philadelphia, opening ice-cream plants in Washington, Boston, New York, Cincinnati, Chicago, and St. Louis. In the 1860s, Fussel was joined by other 'pioneers' in the ice-cream industry, as companies sprouted up all over the United States.

Fussel's absence from the Philadelphia market was probably due to a number of reasons, mainly existing ice-cream enterprises such as Bassetts Ice Cream, founded there in 1861, and still operating out of its original location in the Reading Terminal Market. Another successful Philadelphia ice-cream business was that of the street vendor, William A. Breyer, who in 1866 started operating in the Kensington and Frankford areas of Philadelphia, and by 1882 had opened his first retail store at 2776 Frankford Avenue. His company grew, and in 1896 he had opened the first ice-cream plant on East Somerset. His son, Henry, founded the Breyer's Ice Cream Company in 1908, and in 1924, built the largest ice-cream plant in the world.

Immigration played a significant role in the development of the culinary arts and the ice-cream industry in the United States. After the French Revolution, a small stream of refugee chefs and pastry cooks came to America to open cafés, restaurants, ice-cream parlours, and pastry shops. And, while many immigrants may not have been chefs, there was definitely a French influence on Philadelphia food habits. In 1825, Elias Durand, a French immigrant who had been an apothecary in Napoleon's army, opened a medical supply store that resembled a modern drugstore. Located near Independence Hall, on the corner of Sixth and Chestnut Streets in Philadelphia, Durand's store centered around the selling of sparkling water which caused Durand's establishment to become a social centre and meeting place for the local scientists, physicians, and literati. Another French immigrant to Philadelphia, Eugene Roussel, created the next sensation—he bottled flavored soda water in the late 1830s.

With the advent of the soda fountain, the invention of the ice-cream soda was inevitable. In 1874, at the fiftieth anniversary of the Franklin Institute of Philadelphia, Robert M. Green, a soda fountain concessionaire, ran out of the fresh cream he was mixing with soda, and hoping that no one would notice he replaced it with vanilla ice-cream. But, people did notice, and they loved this new concoction even more than Green's other creations.

Mrs Sarah Tyson Heston Rorer lived in Philadelphia in the late 1800s. As principal of the Philadelphia Cooking School, she not only taught hundreds of young women how to cook, but also wrote numerous cook books. Her *Dainty Dishes for All The Year Round. Recipes for Ice Creams, Water*

Ices, Sherbets, and Other Frozen Desserts (Philadelphia 1890), contained sixty recipes for ice-creams and recalled the original way ice-cream was served, i.e. in dainty portions. Her book was sponsored by North Brothers Manufacturing Company, who were the makers of the Gem Freezer, 'the best [ice-cream freezer] in the world'. A significant aspect of Rorer's cook-book was the twenty-seven recipes for 'Philadelphia Ice Creams', which differ from other recipes in that they contain cream, but not eggs; not even the yolks are utilized. 'Philadelphia' or 'straight cream' ice-creams, lacking eggs were based on older, more traditional English and French recipes. While the French eventually began to rely on egg yolks to create a richer product, what was regarded as an English attachment to ice-cream made without eggs, persisted. Since Philadelphia was primarily settled by people of English descent,—though many French did immigrate to the city also—this would seem to explain the lack of eggs in 'Philadelphia Ice Cream.'

Mrs Rorer's *Dainty Dishes* also included seven recipes for 'Plain Ice Cream', four for 'Neapolitan Ice Creams', six for 'Delmonico Ice Creams', four for 'French Ice Creams', and twelve for 'English Ice Creams', and perhaps these latter recipes were the inspiration for 'Philadelphia Ice Cream' which was made with either a pint of cream or milk, a half a pound of sugar, and no eggs.

The development of ice-cream creations also shows the struggle for imagination in the ice-cream industry. While a recipe for *cornets*, made from a *gaufres* or waffle batter can be found in an eighteenth century cookbook by Menon, it was not until 1896 that Italo Marchiony attempted to patent an edible, sloping-sided cup. His creation failed to catch the public's attention; and Ernest Hamwi, concessionaire of Persian waffles or *zalabia*, was credited with inventing the ice-cream cone during the 1904 St Louis World's Fair. Two Wisconsin natives, George Hallauer and George Giffy, claimed to have created the ice-cream sundae, in the 1890s. Ice-cream novelties, such as the *Eskimo Pie, Popsicle, and Drumstick*, were first marketed in the 1920s, and have since inspired many other creations, as well as allowed for the popularity of the 'hokey-pokey' and the 'Good Humor Man'.

In order to discover what flavours early ice-cream eaters enjoyed, one must rely on the recipes that survive, since little other evidence was recorded. *Thomas Jefferson's Cook Book* provides some evidence of the fruits and flavors available in the United States in the eighteenth and nineteenth centuries, and shows also that Jefferson was influenced by his sojourn in France. Virginia Randolph, Jefferson's granddaughter, recorded family recipes for strawberry, peach, pineapple, and lemon ice-creams, in addition to an elaborate custard-based vanilla ice-cream. Today, vanilla and chocolate ice-creams are respectively the first and second ice-cream flavours in North America. But these flavours were slow in reaching America, because first of all they had to be adopted into

European tastes—which happened after the exploration and settlement of South America—before they 'travelled' to America. Thus, other flavours evolved in America especially due to local products.

Regional ice-cream flavours in the United States today reflect dominant locally-grown products and they elucidate, for example, a Southerner's inclination for pecan, peanut, and peach ice-creams, a Californian's preference for date-nut ice-cream, and a Northerner's craving for cranberry, maple, and cherry ice-creams. Of course, the creation of new and different flavours is an ongoing project of the ice-cream industry, but there are some limits to public taste. Consequently there is a need for 'coherent savors' to entice the consumer, and with a market filled with manufacturers, the concept that competition motivates the creation of novel flavours simply because of their novelty, might explain why the larger ice-cream manufacturers have copied the ideas of those willing to experiment.

Ben Cohen and Jerry Greenfield are two innovators to whom ice-cream manufacturers look for inspiration. Founded in 1978, with a background based on a correspondence course on the making of ice-cream from the Pennsylvania State University, *Ben & Jerry's Homemade* ice-cream shop opened in Burlington, Vermont. Since then, the popularity of *Vermont's Finest All Natural Ice Cream* has grown considerably, due to their 'funky, chunky flavors, made from fresh Vermont milk and cream'. A 1990 list of original ice-cream flavours contains: *Cherry Garcia*, a vanilla ice-cream with cherries and chocolate chunks, *Chunky Monkey*, a banana ice-cream with chocolate chunks, and *Heath Bar Crunch*, toffee candy in vanilla or coffee ice-cream. Along with their *Brownie Bars*, an ice-cream novelty, *Ben & Jerry's* makes yoghurts, light ice-creams, seasonal specialities, and featured flavours. Recently featured flavours—*Tuskegee Chunk*, a peanut butter ice-cream with chocolate chunks, and *Chocolate Chip Cookie Dough*, a vanilla ice-cream with raw chocolate chip dough—are not only favourites of mine, but *Chocolate Chip Cookie Dough* is also at the top of the 'most-popular-in-pints list'—the list which ranks the best selling flavours of *Ben & Jerry's* ice-cream available in stores in pint-sized containers.

In addition to their unique flavours, *Ben & Jerry's* has a mission statement that includes not only a products' mission, but social, economic, and political missions as well. This is evident from some of their product lines, such as *Ben & Jerry's Peace Pops*, and by supporting the organisation *1% for Peace*, whose symbol appears on all of their packages, this company is trying to improve the quality of life of a broad community. It is just such a goal that is also the motivation behind 'Conscious Concoctions'—flavours like *Chocolate Fudge Brownie*, *Rainforest Crunch*, and *Wild Maine Blueberry*—that 'benefit the employment of underskilled persons and the preservation of the rainforest, and support traditional Native American economy'.

From its beginnings as a luxury food item for the elite, to its status as a socially correct product available to the general public, ice-cream

172 Milk and milk products

has never ceased to be a palate pleaser. Whether one chooses an eggless Philadelphian variety, or dares to try *Ben & Jerry's Dastardly Mash*, it can be hard to imagine this creamy treat once began as fruit flavoured snow.

SELECT BIBLIOGRAPHY

The American Way. A History of the Breyer Ice Cream Company, Philadelphia, 1946.
J.-P. Aron, *The Art of Eating in France*, New York, 1975.
Ben & Jerry's 1990 Annual Report, Vermont Ben & Jerry's Homemade Inc. 1991.
C. Child, *Brief on the History and the Present Meaning of the Term Ice Cream*, Washington D.C. 1915.
Confectioners' Journal XXVIII:326 Philadelphia March 1902, 57.
P. Dickson, *The Great American Ice Cream Book*, New York, 1972.
C. Dueker and J. Dueker, *The Old Fashioned Homemade Ice Cream Book*, New York, 1974.
A. Gottschalk, *Histoire de l'alimentation et de la gastronomie depuis la préhistoire jusqu'à nos jours*, I-II. Paris, 1948.
M. A. Hines, G. Marshall, W. W. Weaver, *Larder Invader. Reflections on 3 Centuries of Philadelphia Food and Drink*, Philadelphia, 1986.
H. P. Horry, R. Hooker, *A Colonial Plantation Cookbook*, South Carolina, 1984. *Ice Cream Past and Present*, Washington, 1990.
M. Kimball, *Thomas Jefferson's Cook Book*, Charlottesville, 1976.
S. Krensky, *Scoop after Scoop*, New York, 1986.
L. Lantz, *Old American Kitchenware 1725–1925*, Camden, 1970.
W. Root, R. de Rochemont, *Eating in America*, N. York, 1981.
D. Schwartz, 'Life was sweeter, and more innocent, in our soda days', *Smithsonian Monthly*, 7 July 1986.
M. Simeti, *Pomp and Sustenance. Twenty-five Centuries of Sicilian Food*, New York, 1989.
W. S. Stallings, 'Ice Cream and Water Ices in seventeenth and eighteenth Century England', *Petits Propos Culinaires (PPC)* supplement 3.
R. Tannahill, *Food in History*, New York, 1973.
H.-J. Teuteberg, 'Periods and Turning-Points in the History of European Diet: A Preliminary Outline of Problems and Methods' in Fenton and Kisbán (eds.), *Food in Change: Eating Habits from the Middle Ages to the Present Day*, Edinburgh, 1986.
Mrs. Sarah Tyson Heston Rorer, *Dainty Dishes for All The Year Round. Recipes for Ice Creams, Water Ices, Sherbets, and Other Frozen Desserts*, Philadelphia, 1890.
W. W. Weaver, *35 Receipts from 'The Larder Invaded,'* Philadelphia, 1986.

SUGGESTED READING

R. Agne, *How do Ice Cream Manufacturers in the Philadelphia area determine what flavours they will carry?* Philadelphia, 1948.
R. Briggs, *The New Art of Cookery, According to the Present Practice*, Philadelphia, 1792.
T. J. Brobson, *Sales promotion in the dairy industry in Philadelphia* Philadelphia, 1937.
S. Crumbine and J. Tobey, *The Most Nearly Perfect Food: The Story of Milk*, Baltimore, 1929.
E. David, 'Fromages Glacés and Iced Creams.' PPC 2:23–35; '*Hunt the Ice Cream.*' *PPC*, 1:8–13; 'The Harvest of Cold Months.' PPC, 3:9–16.
D. Dolfie, *A survey of consumer buying and patronage motives and habits for ice cream in a selected Philadelphia district*, Philadelphia, 1950.

R. Harbison, *Collective bargaining of the milk industry in the Philadelphia area*, Philadelphia, 1947.
R. Lenox, *Changing channels of distribution for fluid milk in Philadelphia*, Philadelphia, 1957.
R. Selitzer, *The Dairy Industry in America*, New York, 1976.
C. Wegandt, *Philadelphia Folks*, New York, 1938.

APPENDIX
Glossary of Ice-cream terms

The following are terms and definitions used to distinguish the types of frozen creams and ices. There is a need for differentiation in order clearly to comprehend the regional and ethnic variations of 'ice-cream'. [The glossary is largely based on C. Child, *Brief on the History and the Present Meaning of the Term Ice Cream*, Washington D.C., 1915, with some reference to M. Simeti, *Pomp and Sustenance. Twenty-five Centuries of Sicilian Food*, New York, 1989, and *Ice Cream Past and Present*, Washington, 1990.]

Italian terms:

Gelato: iced 'thing', a general term for frozen confections in Italian.
Granita and *Gramolata*: granular ices made of fruit juices, wines, coffee, etc, boiled with water. The mixture is placed in a pan and scraped down every half an hour to produce a grainy, flaky texture.
Sorbetto: frozen beverage like *granita*, but with a smoother consistency, and more sugar.
Sorbetto can be made with or without cream, and is served semi-frozen.
Gelato di crema: strongly favoured ice-cream, served semi-frozen.
Crema: confection served either frozen or unfrozen, *crema* is used to denote either condition. With *crema*, the 'tendency to cite the practice of foreign nations appears even in Italy, where the making of ices started, as *crema francesse alle mandorle, crema francesse d'albicocche*, 'French almond cream', 'French apricot cream".
Ghiacciato: confection having its own special name, such as *zabaglione ghiacciato*, a mixture of wine, eggs, sugar, and spices, beaten to form a punch, served iced or frozen.
Spongata: frozen cream that is beaten or whipped to a light, frothy texture. This cream was originally layered with *pan di Spagna* or sponge cake, but today refers to ice-cream in a cup. (Simeti, *op. cit.*, 297)

French terms

Glace: an all-inclusive term for ice, though some confections may contain cream.
  Glace, occasionally spelled *glacée*, could be a 'fluid frozen and taken for refreshment, such as *glace à la crème, à la vanille, au citron, au chocolat, etc* . . .'
Glacée: literally 'frozen, . . . when prominence is to be given to a material or mixture used, or to the form used'.
Crème: cream and mixture that may be served either frozen or unfrozen.
Neige: meaning snow, is an ice of soft consistency, 'though not considered exclusively, a water ice, since no special ingredients or manner of freezing seem to be indicated'.

174 Milk and milk products

Fromage: meaning cheese, is a sweet tasting glacé made of milk, cream, egg yolks and sugar, moulded sometimes in the shape of a cheese; it is the predecessor of *bombe glace*.

Bombe: named for the spherical mould, in which two mixtures are frozen in different layers.

Tortue: meaning tart, which was indicative of how the ice-cream was served, i.e. tart shaped.

Cannelon: an ice-cream that was always cylindrically moulded.

Sorbet: derived from the Italian *sorbetto*, it is made without cream, and is similar to *sherbert* or soft cream; it resembled mostly soft ice, or a softly frozen beverage

Mousse: a frozen, frothy, foamy, whipped cream of milk.

Frappé: a partially frozen water ice, with the consistency of *granita*.

Plombière: 'elaborate [ice] made with the addition of whipped cream, nuts, candied fruits, etc.'.

Some of the ambiguous French terms need further clarification. *Glace* is translated as 'water ices' or *eaux delicieuses*, literally 'delicious waters.' *Eaux*, which can have a medicinal interpretation, may not even pertain to water, but implies the juice of fruits or vegetables. The difference between *glaces aux fruits à l'eau*, and *glaces à la crème*, is that the fruit ices are made of sugar, water, and different fruit flavors, while ices of the cream kind are made of cream of milk or milk, eggs, sugar, and different flavors.

English terms were based on French terms:

Ice: related to the French *glacé*.

Cream: a mixture that may be served either frozen or unfrozen, when frozen, 'ice' usually precedes the term cream.

American terms are due to government legislation that regulates the percentages of air, eggs, cream, fat, etc., ensuring that a product conforms to the federal standards:

Frozen custard: ice-cream with 1.4% of egg yolk

Ice milk: 'ice-cream' with a milkfat content between 2% and 7%.

Sherbet: frozen confection with a low percentage of milk, and a higher percentage of sugar and fruit juices.

Sorbet and *water ice*: similar to *sherbet*, but made without milk products and by agitation.

Mellorine: product similar to ice milk, containing vegetable fat or oil.

A Quiescently Frozen Confection: a popsicle, a frozen dessert similar to *water ice* though not agitated, often served on a stick.

Bisque: ice-cream layered with a sponge cake or lady fingers, reminiscent of the original *spongata*.

Frozen novelties: an umbrella term that includes dairy and non-dairy products usually sold by a street vendor; ice-cream, *Eskimo Pies*, ice-cream sandwiches, and *quiescently frozen confections* are all *frozen novelties*.

Ice-cream recipes

Philadelphia Ice Cream

The following recipe is adapted from Mrs Rorer's *Dainty Dishes For All The Year Round*

1 quart of cream $1/2$ pound granulated sugar   1 vanilla bean

Put half of the cream in a double-boiler over the fire, add the sugar, split the bean into halves, scrape out the seeds carefully and add to the hot cream, then throw in the bean, stir it around until the sugar is dissolved. Remove the bean, take the cream from the fire, and when cold add the remainder of the uncooked cream, freeze. (Rorer *op. cit.*, 19).

Vanilla Ice Cream

The following recipe is adapted from *Thomas Jefferson's Cook Book*.
Beat the yolks of 6 eggs until thick and lemon colored. Add, gradually, 1 cup of sugar and a pinch of salt. Bring to a boil 1 quart of cream and pour slowly on the egg mixture. Put in top of double boiler and when it thickens, remove and strain through a fine sieve into a bowl. When cool and 2 teaspoons of vanilla. Freeze, as usual, with one part of salt to three parts of ice. Place in a mould, pack in ice and salt for several hours. For electric refrigerators, follow usual directions, but stir frequently (Kimball, *op. cit.*, 35–36).

Lemon Granita/Granita di Limone

The following recipe is adapted from *Pomp and Sustenance*.

1$^{1}/_{2}$ cups sugar   1 cup freshly squeezed lemon juice   4 cups water

Boil 1 cup of water and the sugar (2 cups for a sweeter granita, 1 cup for a sour granita), make sure the sugar has dissolved. Cool, and stir in 3 cups of water and the lemon juice. Pour the mixture into a shallow pan, and freeze. After half an hour, scrape down the sides and bottom to incorporate the solid with the liquid. Repeat every half hour until slushy, flaky, and firm. Serve. Store in a plastic or glass container; yields 8 to 10. servings (Simeti, *op. cit.*, 284).

16 'Good Humors' and the 'Good Humor Man'

Molly G. Schuchat

For seventy years candy-coated ice-cream individually-wrapped and neighbourhood-delivered, has been a part of American summers. The uniformed ice-cream man, heralding his arrival on the block by ringing bells on his truck has been a local Pied Piper, drawing children from their homes and yards and playground games to line up with the other kids, buy their Good Humors and slowly savour them, sitting on the curb or resting on the nearest emerald lawn, watching the clouds float by in the blue sky above.

This small segment of Americana, recollected by adults and modified for today's children, provides a glimpse of the more general changes in neighbourhoods, in vending, and in choice of foods for impulse purchase in the United States.

ICE-CREAM HISTORY IN GENERAL AND 'GOOD HUMORS' IN PARTICULAR

There has always been ice. But in a state of nature it is a winter material, when people would really prefer it to be available in the summertime. So various humans have been working on saving it, shaving it, flavouring it and transporting it, one way or another, from the always cold mountains to the hot lands below. This has been going on for millennia, from the heydays of Egypt and Babylonia through the Caesars and well beyond. Marco Polo talked of a chilled fruit and milk delicacy from the court of Kublai Khan and later, seventeenth-century Italians brought ice from the hills to make ice-cream.[1]

In the American Colonies, dessert consisting of 'some fine ice cream' was noted in a letter about dinner at the home of William Bladen, Governor of Maryland between 1742 and 1747. Later in that century ice-cream was available at East Coast confectionery shops. Waverley Root suggests that from the names associated with the advertisements, it would appear that ice-cream was first introduced into North America by Italian immigrants. But it could have been the French.[2] In any event, ice-cream was extremely expensive until a hand-cranked ice-cream freezer was invented, leading to the first commercial ice-cream in 1852.

Then, in 1904, a breakthrough in tastes occurred, usually placed at the St Louis World's Fair, where immigrant Ernest a. Hamwi sold *zalabia*, a crisp Persian pastry baked on a flat waffle-iron and served with sugar. An ice-cream concessionaire nearby (ice-cream was very popular at the fair) ran out of dishes for his customers and Hamwi rolled one of his wafers into a cornucopia, cooled it down and put a scoop of ice-cream in its mouth. A great story, but patent office evidence suggests that it was Italo Marchiony, an Italian immigrant, who began selling cones as early as 1896 and was granted a patent on his cone mould just before that fair opened.

Whoever started it, Waverley Root reminds us that Americans like sweetness, eating standing up, and hand-held food that can be consumed rapidly. By 1924, two hundred and forty-five million ice-cream cones were sold annually. Today we can also note that the ice-cream cone is the perfect ecological package.

Christopher Nelson, who dipped a bar of vanilla ice-cream into chocolate, died this year[3] at the age of ninety-eight, but it was seventy-two years ago that he invented the Eskimo Pie. Legend has it that a boy came into his drug store and anguished over the decision between a chocolate bar and a dish of ice-cream when he could only afford one of them. So Nelson, yet another ingenious immigrant, not trained as an engineer but a natural inventor, tried to oblige. His 'I-scream bar', with the name changed to Eskimo Pie, was an immediate success.

My colleague Mary Hamill's father was in at that creation. He was working his way through medical school as a summertime pharmacist in Onowa, Iowa, where Nelson was trying to meld candy and ice-cream. The problem was that the chocolate would break when frozen against the ice-cream. The soon-to-be-doctor suggested cocoa butter (he was familiar with its properties from more medical usage) and that worked out. Nelson offered him a partnership in his invention, but my friend's father wanted to spend his money on medical school. They remained lifelong friends and 'he was like Santa Claus to me and my brothers', reminisces Mary. He gave the family a food locker in the Detroit plant, where Nelson had pioneered the use of dry ice to keep pies solid. Dry ice was a never-ending source of great fun. They took it on picnics, watched it steam in the water and do magic tricks! That, too, is now *passé* with the prevalence of freezers!

Harry Burt, a confectioner in Youngstown marketed a lollipop called the 'Good Humor Sucker', 'Good Humor' being a name that expressed his belief that the humours of the mind were regulated by those of the palate. Burt was always looking for new ideas, so as soon as he heard about the Eskimo Pie created in neighbouring Iowa, he tried to repeat it. He tested his new one on his daughter who loved the taste but thought the bar was 'too messy', and suggested he put it on a stick like his famous lollipop. Lo! the bond formed by ice crystals interlocking with the wood handle worked! Burt applied for a patent on the 'Good Humor Ice Cream Sucker'.[4]

When Burt died in 1926 the rights to his patent were acquired by a group of Cleveland, Ohio businessmen, who formed the Good Humor Corporation of America, to sell franchises. Tom Brimer, in Detroit, bought one and soon went on to Chicago. In 1929 some Chicago mobsters asked $5,000 for protection . . . or else. These were the same Chicago gangsters who developed so blatantly because of the illegality of liquor during the Prohibition era in the United States.

Brimer ignored their demand but increased the insurance on his equipment. Quite soon, eight of his Good Humor Trucks were blown up and this was a front page story. Sales shot up. By the end of the year (right after the Wall Street Stock Market crash) Brimer paid stockholders a twenty-five per cent dividend! . . . and then bought a seventy-five per cent interest in the parent Cleveland company. Although an American poet (Ogden Nash) wrote that 'candy is dandy but liquor is quicker', obviously candy and ice-cream were more available during Prohibition (1918–32), while in the great Depression of the 1930s, a cheap treat like a Good Humor was very welcome.

The Company was a tightly run franchise operation, requiring white uniformed troops complete with shiny Sam Browne belts and a rigorous way of behaving, taught from a manual, *Making Good with Good Humor*. In California, the trucks were parked outside Hollywood radio stations and movie studios, which made for many talk shows and photo opportunities. The trucks are in over one hundred motion pictures from the 1930s and 1940s and Columbia Pictures even made a film, *The Good Humor Man*.[5]

Post-World War II suburban growth to accommodate the baby boom, was also a boon to the Ice-Cream Man's spread to new routes. That, together with the development of supermarkets and the sale of ice-cream novelties by the box from their big freezers, converged so that by 1963, thirty per cent of the entire ice-cream market was novelties, including sherbes, ices and other innovations, all individually wrapped, although sometimes sold in a six-pack.

The high point of the ice-cream market was reached in 1984, when Americans ate 18.5 pounds of ice-cream *per capita*. Per capita consumption has since slumped to less than 15.7 pounds in 1990.[6]

In the late 1980s came a children's book about the marriage of the Zalabia pastry with vanilla ice-cream at the St Louis World Fair.[7] This volume may have been meant to spur would-be inventors, or it could have been an entry in the multicultural praise era of American life. Since the ice-cream quickly softens, it might instead be a last remnant of the melting-pot theory. Or it might even be a public relations-related push to sell more ice-cream products, like the recurring newspaper spring sitings of the Ice-Cream Man in many newspapers around the country.[8]

NOW GROWN-UP CUSTOMERS

Vera, a former Philadelphian, recalls that they did not have Good Humors in her part of town in the middle 1930s. However, she and the other children were often sent to the corner grocery store after dinner's main course to buy ice-cream for dessert. But after moving to Washington late in the decade, Vera would sit out on the grass with a girl-friend after dinner to wait for the announcing bell, rush in to ask for money, and then buy Good Humors. They still could not store ice-cream in non-existent freezers!

Bob lived in Baltimore in the 1930s and there were Good Humor trucks in his neighbourhood, he recalls. But both his parents worked, and he and his brother were busy in the afternoon with homework and making dinner. They usually ate fairly late, so he did not get Good Humors either in the afternoon or in the early evening.

Ron said where he grew up the Good Humor man did not drive around the neighbourhood, but met his customers in one central location in suburban Long Island. He allowed charge accounts, paid by the mothers at the weekend. He knew all the kids and their families and worked in a nearby restaurant in the winter.

Jim recalls that across the country in San Diego in the 1950s, in a not-at-all rich neighbourhood, one would hold the truck while everyone else ran home for money. The Good Humor man did not dawdle there, and if no one was waiting to buy, the truck rushed on. But listening for its arrival and dividing the 'work', kept the kids socially together.

At about the same time, in a very upper class New York City suburb, Louise also loved the coming of The Good Humor Man. He provided her and her friends, who had said goodbye to each other after school, had gone home and then talked to each other on the telephone for the rest of the afternoon, one last opportunity to regroup again by his truck after dinner.

In the small towns around the country there were few Ice-Cream Men in trucks, but some on bikes with cold boxes attached. In many rural areas there were open wagons as well as bicycles, but neither carried ice-cream novelties, only fruit, candy and other non-meltable snacks. These peddlers were not in uniforms of any sort. But all had their distinctive bells, horns or musical calls, not just the Ice-Cream Men.

A Good Humor manager, Dick, who started working in the 1940s and retired in 1991, reminisced about his former days selling from a truck. He recalled some of the many dog stories. There were a couple of dogs on Dick's routes. One route was on a charge account system. He would take the top off of a dixie cup pint, and this particular dog (that had come as a response to the bell, leaving the other dogs with which he was roaming around) would collect the cup and trot up to the porch with it. Another one had money in a pouch on his dog collar.

Back when Dick was driving a Good Humor Truck, they often went through the alleys between the streets, because they were so much safer for the kids to come out to. 'But there aren't any alleys any more and the streets really are too dangerous, with all the cars all the time'.

Dick also had a problem when he followed a route previously run by a left-handed man; he had to teach the kids to line up in the other direction. 'You have to teach the kids to line up anyway, because it's so much harder and takes longer to pick and choose among a crowd standing around than in a line. But they learn; kids today are no different from the past.'

Oscar Hijuelos wrote the 1990 prize-winning novel, *The Mambo Kings Play Songs of Love*, about the hard life and broken dreams of refugees from Cuba in the 1950s. His subjects were after-work musicians and their strong dance rhythms and the dance halls that enlivened the lives of the poor in East Harlem, New York City. One of the heroes, paying court to a young woman, goes with her on a lazy afternoon to the park by the Hudson River, near Columbia University:

> they went to sit on a nice grassy hill ... children were playing softball in a field and pretty college girls in Bermuda shorts and white tennis shoes were spread out on blankets studying their books. The sun up in the sky, a buzz of insects in the air, boats and barges passing on the Hudson river ... Then a 'tling-tling', a Good Humor ice-cream man in his squat white truck. Nestor walked over and came back with two ice siccle pops, his strawberry, hers orange, and they ate them, running with sweet liquid, and lay back. She was so happy because it was a beautiful day and they were in love ...'[9]

Obviously, the symbolism of the Good Humor Man and his truck are very meaningful to the author, who included in his writing other similar scenes, in other parks, in later years.

The combination of the lazy summer afternoon, and the sound and sight and taste of the Good Humor Truck and ice-cream, both reflect, and are included in, the innocence of children, nature, and the sunny days on the hill—a dream, or reality—but certainly the memories of childhood for many of us, not just for Cuban immigrants having a brief time out from the hustling and squalor of their everyday lives. Today, special occasions, such as weddings or elaborate summer parties, often include a Good Humor Truck hired to hand out specialities to the guests. The truck, the bell, the chocolate or nut, or whatever covered bars, return us, even briefly, to innocence, to fun, to playtime of former years.

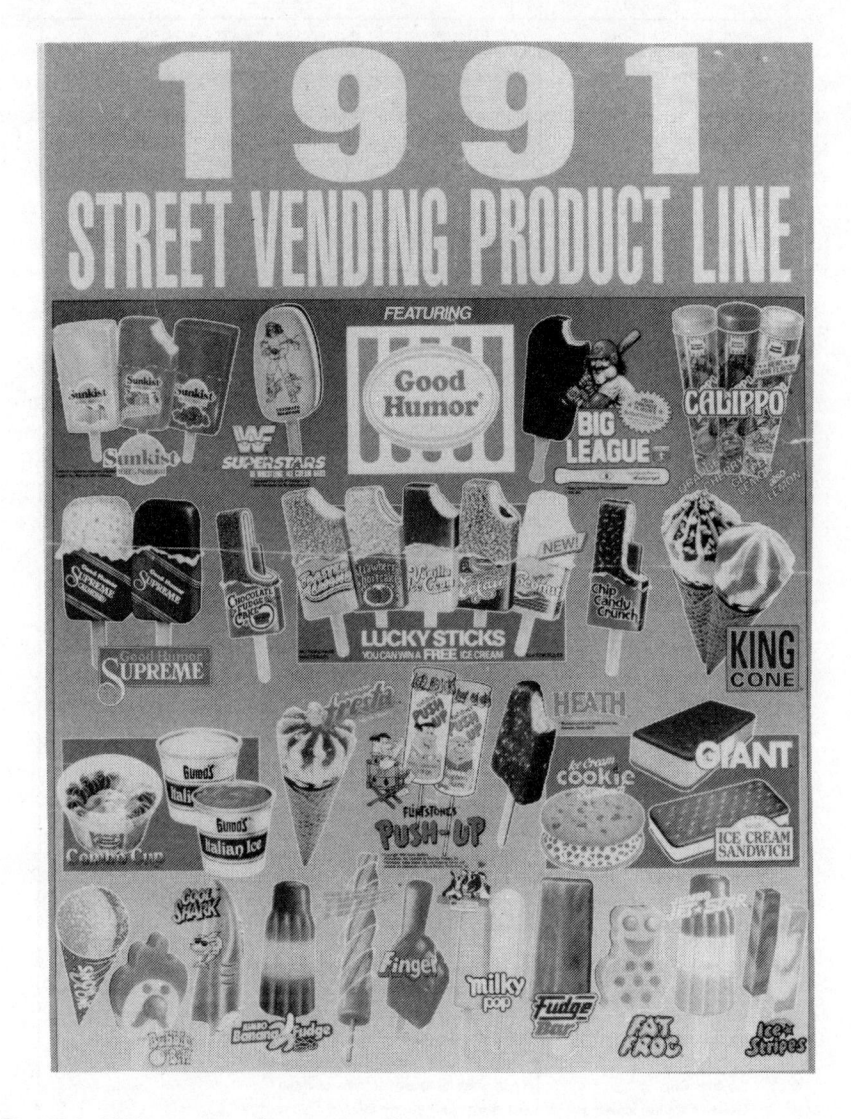

Fig. 16.1. 1991 street vending product line poster

TODAY'S CHILDREN

Back in Washington today, the scene at a neighbourhood swimming pool is described by the mother of small children. 'No one ever told the Good Humor man to come there, but every hour on the hour, like clockwork, he appears, and all those children behind the pool-house, and leaping and

Milk and milk products

Product	Description
King Cone Van/Choc	Vanilla & chocolate ice cream in a crisp sugar cone with chocolate topping and peanuts.
King Cone Strawberry	Vanilla ice cream with strawberry fruit ripple in a crisp sugar cone.
Fresta Mixed Berry Cone	Cherry and raspberry frozen yogurt with strawberry pieces and strawberry sauce in a crisp whole wheat cone. 96% Fat Free.
Supreme-Dark	Premium vanilla ice cream dipped in rich dark chocolate.
Supreme-Milk	Premium vanilla ice cream dipped in smooth milk chocolate.
Supreme-Almond	Premium vanilla ice cream dipped in a special almond coating.
SUNKIST	
Orange Juice	All natural orange juice fruit ice.
Lemonade	All natural lemonade fruit ice.
Wild Berry	All natural cherry and raspberry fruit ice with strawberry pieces.
Orange & Cream	All natural orange fruit ice with vanilla ice cream center.
CALIPPO	
Grape/Lemon	100% Natural grape and lemon fruit ices.
Lemon	100% Natural lemon fruit ice.
Orange	100% Natural orange fruit ice.
Cherry	100% Natural cherry fruit ice.
GOOD HUMOR LUCKY STICKS	
Vanilla	Vanilla ice cream in a dark chocolate coating.
Toasted Almond	Vanilla ice cream in a special almond crunch coating with almond pieces.
Strawberry Shortcake	A strawberry sauce layered through vanilla ice cream coated with our special shortcake.
Chocolate Eclair	Vanilla ice cream with pockets of rich chocolate crunch coating.
Coconut	Vanilla ice cream with coconut flake coating.
GOOD HUMOR - REGULAR	
Chip Candy Crunch	Chocolate chip ice cream with a chocolate candy center and chocolate crunch coating.
Chocolate Fudge Cake	A chocolate candy center surrounded by layers of vanilla ice cream and chocolate fudge with a chocolate cake coating.
Cool Shark	Natural grape flavor with soft white lemon center.
Jumbo Jet Star	Combination of raspberry, cherry and lemon ices.
Jumbo Banana Fudge	Banana and chocolate flavored fudge.
Fudge Bar	Real chocolate fudge on a stick.
WWF Bar	Vanilla ice cream with chocolate coating and printed cookie.
Heath Bar	Toffee ice cream with real Heath milk chocolate.
Flintstone Push Up-Orange	Orange Sherbet.
Flintstone Push Up-Raspberry	Raspberry Sherbet.
Giant Neapolitan Sandwich	Strawberry, vanilla & chocolate ice cream between two chocolaty wafers.
Vanilla Sandwich	Two chocolaty wafers with vanilla ice cream.
Cookie Sandwich	Two chocolate chip cookies around vanilla ice cream covered in chocolate chips.
Big League	Vanilla ice cream dipped in chocolate with crisped rice.
Twister	Orange and lemon flavored water ice.
Combo Cup	Combination vanilla and chocolate ice cream.
Guido's Italian Ice-Cherry	Refreshing cherry flavored ice.
Guido's Italian Ice-Lemonade	Refreshing lemon flavored ice.
Ice Stripes-Grape/Lemon	Mixed grape/lemon flavored water ice.
Ice Stripes-Cherry/Orange	Mixed cherry/orange flavored water ice.
Milky Pop-Vanilla & Strawberry mixed	100% natural vanilla and strawberry ice milk in a mixed case; "No artificial ingredients".
Fat Frog	Green vanilla ice cream, chocolate candy eyes and a chocolate back.
Bubble O'Bill	Chocolate and vanilla ice cream, chocolate coated back, bubble-gumball nose.
Snow Cone	3 big refreshing fruit flavors in one big crushed ice cone.
Finger	Strawberry flavored ice in a finger shape.

Fig. 16.2. 1991 street vending product descriptions

screaming around the pool, hear the bell and rush out, crowd around, buy, and sit on the hill (no food allowed in pool area) with their concoctions . . . until the next hour.'

Elsewhere in the Washington suburbs, every afternoon in spring and fall, a Good Humor Truck parks by school-yards. While the children assemble by the home-bound school buses, there is a 'run'. Designated children rush to the truck to buy ice-cream and candy for the rest of their friends on the bus, for a snack during the ride home. Candy is more popular, but the mothers prefer the ice-cream novelties because 'they are more nutritious'.

At factory gates and by office buildings, too, the Good Humor Trucks are available when the shift ends or during lunch-hour.

At a small, historically Black college in the deep South, students included some from the northern big cities, as well as the rural areas of Mississippi. Asked about the Ice-Cream Man, their faces lit up and everyone wanted to speak at once. Chris said that they only recently got an ice-cream truck in his town, although they always have had a candyman, and open-backed fruit trucks, from which they swiped fruit. They cannot get at the Good Humors, since the truck is all enclosed. They ask someone at home (or at least their little brothers and sisters do) for the money each time and are often told 'can't have ice-cream from the ice-cream man until you've finished your greens. . .'. The Ice-Cream Man has his bell and his songs and knows his customers. He always has their favourites out for each of them.

The ice-cream truck drives to the front of the College administration building in this country town, but he sells only in boxes of twenty-four, which room-mates buy and store in a dorm-room refrigerator. Since most of them do not have cars, it is more convenient to have the truck come to them, than for them to hike the quarter of a mile to a convenience store, even though that would be less expensive. And even though they buy in quantity and keep them in their rooms, now they, too, think of the Ice-Cream Man as part of playtime, summertime, friends' time.

However, their stories of back home and the Ice-Cream Man did not always reflect this feeling.

In Aaron's Philadelphia neighbourhood, the Good Humor Man also knows everyone, and gives the mothers who are working and not at home when he swings by, their bills at end of week. Little kids run after the truck, but teenagers do not run after Good Humor Men; they stroll, and so sometimes they do not get to him in time!

A couple of Chicago youngsters said they always had to ask for money, because they were not supposed to carry it around, since it might get taken from them. They talked, too, about bike 'trucks', that could get held up from time to time. A lawyer elsewhere suggests that Good Humor Trucks are 'sitting ducks', waiting to be robbed; 'it's a great deal, instead of having to go to the bank, the bank comes to you'.

184 Milk and milk products

In some ghetto neighbourhoods, one young tough guy said, the Ice-Cream Man sold ice-cream to kids and dope to the teenagers. Sometimes the trucks were robbed. And sometimes on street corners in Chicago, frozen novelties are sold from boxes (stolen?), quick, before they melt; it is cheaper than from the Ice-Cream Man's truck!

A few of the students were not allowed to buy ice-cream or anything from the vendors, because of the oft-repeated story of the man who sold or gave ice-cream, played with the little kids, and then shot them. 'Don't you know about him?', asked a student from Kansas City, agreeing with one from Detroit.

The following comment recently appeared in the *Washington Post*, 'Letters to the Editors column:

> Walking home one Sunday evening (in the heated, humming heart of Adams Morgan) I saw the familiar silhouette of the Good Humor Truck. As I passed and looked at the queue of kids clamoring to buy their favorite ice-cream, I was horrified to see that the window I remembered from my childhood as being open to customers was covered by iron bars and chicken wire. When things have gotten so bad that the minimal cash a Good Humor driver has must be protected by bars and wire, the city is in a sorry state. It's no longer possible to enjoy a simple summer pleasure like ice-cream.[10]

VENDOR'S VIEW

Dick, the retired Good Humor manager, said that he was working in a restaurant in 1945. A Good Humor Man would come in for meals and told Dick that he was making $200 a week. Dick was making $60 and his meals at the restaurant. After he actually saw the customer's pay-check, he changed jobs. Eventually he moved on from being a vendor to having an indoor supervisory job. People still become Good Humor Men the same way; 'People come through other people'.

But today 'mostly they're lazy, they don't finish the routes but do so much and quit, not like years ago'. Dick worked long days in the summer from 6.30 am to 9.00 pm, after supper and before dark, and in winter after school until 5.00 pm. He never took a day off between April and October. But still, 'you are your own boss, no one is ordering you when and where to do everything, no one follows you around'.

Recently one ten-year-old called the Franchise office because the Good Humor Man had not come at 4.00 pm as he usually did. The caller had rounded up a group of friends, with money ready, like he usually did. And the Good Humor Man had not come the next day, either. It was now 5.30 pm and the driver not there yet! 'Where is the truck?'.

'That was a resourceful kid', said Dick, 'and kids are the best customers in the world. They know what they want and they go after it; if they like it they keep at it, but their tastes change . . . great faddists . . . some things come in and are hot and then are gone'.

Good Humor's marketers try to guess what the children want. One company was very successful with ninja turtles, but that fad is fading. A novelty that has lasted is a Good Humor shaped as a frog with eyes made of M and M-multi-coloured, candy chocolate bits; faces of any kind on the ice-creams are always very good sellers.

'And if it's something that "grosses out" or disgusts the parents, that's even better', suggested the franchise owner on another occasion. Now the franchise owner might disagree with Dick's evaluation of the continuity of children's behaviour. Or at least he thinks parents are different, in that they do not supervise their kids or have any one else do it, either . . . which leads to lawyers, who together with the insurance companies, are the bane of everyone's existence! After a speeding auto hit a kid who was going across a street to buy Good Humors and the child got a broken leg, the parents sued the car owner, who was at fault. But they also sued the vendor and then the Franchise and the Good Humor Company, for two million dollars each! That was because the vendor did not get out of the truck and cross the street to sell to the kid. 'How can the vendor get out of his truck at stops, he'd never get through a route that way. So why didn't they sue the cow for producing the milk, and/or the farmer for producing and selling it in the first place?'.[11]

There are not many trucks riding through the upscale neighbourhoods, because there the children are not allowed to play in the streets, and they have big backyards. The children cannot hear the bell from the backyard and do not rush out. Also vendors need houses closer together than homes in suburbs, since their business is dependent on volume, on the nickels and dimes that add up with hard work.

Moreover, in summer, the 'rich' kids are in camp, while during the school-year, after-school activities away from home keep them from becoming regular customers. They do a good business right after school, but then the children scatter to their various activities.

In housing complexes and other apartment developments, the children still play in the interior streets and have fewer after-school activities. But more and more, the managers are telling the vendors that the trucks cannot come any more because there is just too much mess all over the streets; people just do not use waste baskets any more.

More and more zoning ordinances are limiting the drivers' routes. A recent *Wall Street Journal* article warned that street vendors are facing ever-tougher curbs. Earlier, a doctoral dissertation discussed the vending situation in Chicago in the 1980s, the hardships for, and exploitation of, today's vendors, as well as the business and other regulations that encroach on them.[12]

Street vending and peddling have been a traditional entry into American life for immigrants through the centuries. Vending via an ice-cream truck, is certainly a step-up financially from selling via a back-pack, or from push-carts and street stands. For one thing, it requires a driving licence and other technical skills. Vendors usually start by renting a truck from someone else, or working for an owner. Today, most Good Humor vendors in Washington are from Iran, Thailand, West Africa and Central America. There are several West-African vendors, all of whom are excellent sales persons, and while there are a few American Blacks, there are no African-American women vendors.

Since the vendors are independent contractors they can run their own hours, and the franchiser has no dismissal recourse. But he can offer incentives of more discount for greater sales volume and end-of-the-year bonuses. But if the vendor is from a country where $500 (a week's profit in the high season) may seem like heaven, he may not work that sixth or seventh day because he has so much money in his pockets! Those who work hard for the seven months major season, either take other jobs in winter, or quite often go to their home countries to visit with all that money ... Out of the whole group, there are still about three per cent 'older' white men, over forty-five years who have been Ice-Cream men all their lives and like the customer contact. Generally, the vendors have a great time on the parking lot, with the colourful interplay of all these cultures and little acrimony between the different ethnicities.

SOME DIRECTIONS OR CONCLUSIONS

Styles in ice-cream novelties change swiftly, following the same paths as other fads. In the stores they vie with high-priced and other low-priced items in the ubiquitous freezer cases. In an article on ice-cream in 1981, a food writer was quoted, 'I loved Good Humors because they had chocolate shells ... as I grew older, I dared more chocolate concoctions ... and now buy a six-pack of something called a 31-karat bar, at the Baskin Robbins store on the way home from work and eat them all up by the 11 o'clock news'.[13]

Some years back, the Good Humor Company brought out a chilli con carne ice-cream bar. It failed resoundingly. Perhaps they should try again now that America (and the world) is enmeshed in a hot chilli love affair. The South Asians eat yoghurt with their chillies and curries, but since Americans have this profound sweet tooth, chilli-coated ice-cream's time may finally have come ... except that all the scientific eating we are being offered eschews both ice-*cream* and *milk*. What used to be the athletic food of choice, the glass of milk and all the other products, are currently being replaced. Training table-talk is more likely to revolve around the difficulties of digesting ice-cream in any form, rather than the delights of a favourite flavour, or even mourning the innocent days of childhood lining up for the Good Humors on hot summer afternoons.[14]

Everywhere, cold drinks are sipped through straws that emerge from holes in the top of paper drink-cups, or new sleek thermoses with their very own private straws. Cokes, and other ubiquitous soft drinks, are imbibed while walking, sitting in a car, or on a bus, or in the movies, or at the ball game—almost anywhere and almost anytime. They are much less messy than Good Humors or popsicles for the consumer, although neither are nutritious, and the paper-cup ones are not ecologically sound.

Good Humors as a 'home delivery' convenience, and as a fond reminder of childhood's minor high points, are losing their significance, and as they do, we may lose the street rituals of the summer comings-together that they encouraged. However, we may not be diminishing childhood in that way, even though we may be adding to the loss of a sense of neighbourhood as we have known it.

We yearn for the 'age of innocence' that was and is no more, while our current horror tales regarding outsiders reflect our many fears of life itself. We have less belief in our traditional values, and so we say the values have disappeared, not that *we* have changed. It has been our myth that children in a blessed age, before the rats invaded Hamelin, lived in peace and richness and tranquillity. We now think that our children's lives are threatened, because we are aware, finally, that most children's lives all over the world are, and have always been, threatened. We have been obliged to leave aside our cherished myth of an 'age of innocence' and face the realities of modern life.

But the rituals, and the friendships that children develop here and everywhere, with or without Good Humor Trucks, do transfer to areas like pools and other primarily non-market places, where children still gather together.

And on a far more mundane level, perhaps the fear of fat that now hovers over daily food habits, will act to revive ice-cream, in dishes and on sticks, as treats to be treasured and remembered, not as quick fixes to be gobbled indiscriminately and too easily.

REFERENCES

1 Early history of ice and ice-cream can be found in R. Tannahill, *Food in History*, New York, 1973, and in L.W. Haynes, *An Analysis of the Ice Cream Industry as an Outlet for Dairy Products* (Doctoral dissertation University of Illinois at Chicago, 1988.
2 W. Root and R. de Rochemont, *Eating in America*, New York, 1976.
3 Obituary in the *Washington Post*, March 11, 1992.
4 P. Dickson, *The Great American Ice Cream Book* New York, 1977, esp., 77.
5 See Dickson, *op. cit.*, for more detail.
6 United States Department of Agriculture: 'Per Capita consumption of Major Food Commodities', *Agricultural Statistics*, 1991.
7 S. Kroll, *The Hokey Pokey Man*, New York, 1989.
8 For example, there was a story on the metropolitan Washington Good Humor Franchise in June 1991 in a series of suburban newspapers.
9 O. Hijuelos, *The Mambo Kings Play Songs of Love*, New York, 1990, 88.
10 S. Crawford, 'Iron Bars and Ice Cream Bars', *Washington Post*, August 8, 1992.

188 *Milk and milk products*

11 The lawyer tells me that argument is termed 'too remote' by the courts. Obviously it *has* been tried.
12 C.Eastwood, *A Study of the Regulation of Chicago's Street Vendors*, (Doctoral dissertation University of Illinois at Chicago, 1988.
13 J. Skow, 'They All Scream for Ice Cream', *Time*, August 10, 1981, 52–58.
14 See related material in Tracey De Crosta, 'Summertime, Promise or Pitfall?', *Bicycling Magazine*, 1981.

Part V
Milk and Folk Belief

17 Milk and folk belief: with examples from Sweden

Anders Salomonsson

Since time immemorial milk has been one of humanity's most important foods, and since the Neolithic period, people have raised domesticated cattle partially for the sake of milk. It was not only cows which were kept for milk purposes, however; sheep and goats were also raised for their milk in many places. In the following analysis, however, I shall concentrate upon the milch cows.

In pre-industrial society, both the natural environment and the home economy were factors determining whether one kept cows. Cows required summer pasture and winter fodder, and since the forest was often utilised as pasture, and the meadowlands to provide hay for winter feed, animal husbandry was less important in areas which were predominantly arable in nature. A certain minimum amount of land was necessary to keep one or more cows, and many people in society had holdings under that limit, or no land at all. As a rule, crofters, craftsmen, and fishermen were forced to buy or trade for milk and milk products from the larger farms.

Ordinary sweet milk was seldom drunk fresh, and this was especially the case in households which did not have their own milk supply. For example, in pronounced tillage areas, the common mealtime beverage was usually a sort of beer which was brewed from grain malt. In Scania, in the most southerly part of Sweden, this beer was often even used with porridge instead of milk. But the most important reason why fresh milk was seldom drunk was that it did not keep well. Thus soured milk was the more usual beverage at the dining table. But it was as different kinds of cheese that milk was especially significant in the common peasant diet. Thus cheese-making can be seen as a form of milk preservation, a way to give the important foodstuff a longer life. Cheese was also used as a form of taxation and tithe, and it was an important item of barter characterisinga practically moneyless economy.

THE THREATENED 'MILK-LUCK' (*MJÖLJLYCKA*)

Since milk was an important part of the common diet, access to milk and milk products was of vital importance. However, for many reasons the supply could be uncertain, and fluctuations in milk were explained by saying that some farms had good 'milk-luck' while others had bad 'milk-luck'. For example, the cows on certain farms might go dry earlier in the pre-winter period and begin to produce milk later in the spring; in some places the milk-yield stopped for seemingly inexplicable reasons during the height of the milking season, and so on. 'Milk-luck,' or conversely milk misfortune, was often tied to a farm's matron or the women involved in the farm work, since all the handling of the cows as well as the milk was specifically women's work.

Like all inexplicable and uncertain phenomena, 'milk-luck' has attracted to itself a multitude of different kinds of beliefs and customs; fishing and hunting luck are similar examples of this phenomenon, as is the luck connected with the fermentation of beer. When the farm matron or the maid poured the yeast into her beer brew, she was supposed to scream loudly, or violently wave her skirt, in order to scare away evil powers which could influence the important process of fermentation. It was also important not to allow any strangers into the brewing house or to look at the beer barrels before the fermentation process was well under way. The stranger could of course have 'the evil eye' and thus the potential to destroy the sensitive brewer's wort.

In relation to cows and milk, a large body of rules, rituals and forms of good advice have been recorded and stored in our archives of tradition. One way to obtain a good milk-yield was to breathe into the container before initiating the milking. This also ensured that the milk would not go sour too quickly. Another trick was to throw a few grains of salt, or a few drops of cold fresh water, into the milk cask. A glowing coal could also be put into the milk, or the milk could be passed over a fire three times. When one carried an open milk container out of doors, from the pasture or from the barn, for example, one was supposed to cover it with an apron. This was especially important in relation to milk which was sold or traded to another person. By observing these rules, people attempted to ensure that the purchaser did not later use the milk as a medium to steal the power or substance from the milk itself, or from the cow which had produced the milk. It was consequently one means of maintaining and protecting one's own 'milk-luck'.

This is a belief which not only marked the boundaries between the rational and the irrational, but also marked a social boundary and a hierarchical relationship between milk vendors and milk purchasers, that is to say, between rich cow owners and the poor who lacked cattle, or between established milk givers, and suspect milk takers. The persons who were forced to purchase milk were not secure farmers or landowners like those who could spare milk and sell it to others. They were, therefore, outsiders in a proprietary community, and were considered dangerous since they were believed to constitute a possible threat to proprietorship, to what was established and paternally inherited. Thus in order to protect one's property and luck, one furnished oneself with remedies and protective mechanisms of the type described here.

The common folk conception of the world was handy in this connection since there existed an entire set of ideas about how one should handle potentially dangerous and critical situations. The same or similar remedies were also used to avoid disease, poverty, and different types of misfortune.

Still another very common method of holding misfortune or evil powers at bay was to use iron or steel. Here are some examples from our archival notes: 'A knife was placed under the threshold of the barn for the cow to step over when she was to be milked'; 'When one milked a cow with the intention of selling the milk, one should have a pair of scissors lying in the bottom of the cask'; one should lead the cow over a scythe or make the sign of the cross over the container of milk with a knife. It was especially important to observe these rules the first time ever that one milked a cow, or the first time that the cow was milked after calving.

MILK STEALING

No matter how one tried to protect oneself, variations in milk luck occurred between farms and between the women. Temporary imbalances in one's milk luck could perhaps be accepted, but one was forced to have more enduring and obvious 'injustices' explained and remedied. The peasants lived in a society which was built on balance or equilibrium, one in which each and everyone kept a close eye on neighbours. Social control was effective, and not only was the neighbours' moral conduct watched, but one carefully noted their sowing and harvesting, the diligence with which they worked and the yields from their labour, as well as the degrees of prosperity and misery which accrued to them. Deviations from the norm were immediately observed; they were discussed, and explanations for them were sought since they were seen as a threat to both society collectively, and to the individuals within society.

Widespread in Swedish—and European—popular belief, was the idea that one could, with the help of certain tricks, steal milk from another person's cows without actually having to be present oneself. However, one needed assistance to accomplish this feat and the necessary help was believed to be derived from the Devil. The Devil cold be invoked by means of certain rituals, and when contact was established, one sold one's soul to him in return for the necessary help. This was a dreadful step to take because as a result of it one placed oneself in the Devil's power and promised to burn with him in hell after death. One can speculate as to how many people, who were aware of the consequences, really tried to make such a contract with the Devil, and what driving force led them to this course of action. No doubt the majority of all alleged contracts with the Devil were just an evil figment of neighbours' terror-stricken or malignant imaginations; but perhaps some people were willing to take the risk; one can only wonder why?

According to popular belief one could achieve the most diverse benefits from selling one's soul to the Devil, but in our case it had to do with obtaining ability to steal milk from those who lived in the area. This theft could be performed in different ways. One of the simplest ways was, after having made a pact with the Devil, to strip naked and stand on a high place (preferably a pile of manure) and call out, 'The milk is mine as far as my cry can be heard', or some such words. Almost without exception it was said to be a woman who performed the act, and preferably it should be carried out on a Thursday night, and one could ensure that the invocation remained valid for an entire year by performing the ritual on the Thursday before Easter.

In terms of popular belief, nakedness is, of course, a challenge to established values, and also a shame associated with the Devil. The manure heap is hardly a place that one seeks out for honest purposes, and the eve of Friday was the commencement of the Friday fast (and in the case of Holy Thursday, the night before Good Friday), and thus constituted an appropriate time for a special attack on the Christian Church's generally accepted norms of order and decency in society. The above ideas, and also that one could, under certain conditions, usurp power from the area that one can reach with one's voice, occur in connection with other popular beliefs.

According to certain archival notes, performing this latter ritual was enough to fill one's own containers with the milk of those who lived in the area; however, complementary measures were also usually believed to be necessary. These could involve producing a *mjölkhare* ('milk hare') or a *bjära*. The 'ilk hare,' best known in southern Sweden, was said to look like a little woolly or soft animal, something like a hare. It was produced from weaving together sticks, and perhaps some wool, or skin, as well as clipped-off finger nails, or human hair. Then as one dropped a few drops of blood over the creation (preferably from the left little finger) one read an invocation of the following type:

Jag har gett dig blod	I have given you blood.
Fan skall ge dig mod.	The devil will give you courage.
Du ska för mig på jorden rinna.	You will run for me on earth.
Jag skall för dig i helvete brinna.	I will burn for you in hell.

After that, the 'milk hare' was sent off to the neighbours' milk sheds, drank up all their milk, and then ran off home and vomitted up the milk into the thief's own casks. According to certain archival notes, the 'milk hare' might also suckle directly from the cow's teats, and in that case it caused small scars and flecks on the udder. It was useless to shoot at a milk hare with an ordinary bullet; a silver bullet was needed to bring the beast down, whereupon it would collapse into a pile of sticks and straw.

Similar ideas about milk stealing are noted from northern Sweden; however the milk thief was there called *Bjära* (compare the Swedish *bära* meaning 'to bear' or 'to carry'). The shape of this milk-stealing creature was different from the southern one and in the tales some tradition bearers liken it to a ball which rolled forth over the ground as fast as lightning.

Furthermore, there was another means of stealing a neighbour's milk—by symbolically milking an object associated with the neighbour's cows; for example, one could stick a knife into a wall and obtain milk out of it, or one could obtain milk by means of a piece of fencing or a suspender; even trouser legs, a rope, a broom, and the leg of a chair, could be used according to certain archival notes. It is said that a female milk stealer's powers were so great that she could milk blood from the neighbour's cow if she did not stop in time; this could account for the death or sickness of an animal.

Ideas about milk stealing creatures are clearly very old; medieval church paintings attest to this. When the persecution of witches got under way in Sweden in the seventeenth century, these old ideas coalesced with witchcraft hysteria, and one of the most frequent accusations against women branded as witches was that they had stolen milk with the Devil's help. In the preserved witch trial protocols testimony is given about how the witch worked her magic, just as there are confessions—often obtained through duress—of milk stealing from the accused woman.

MILK-LUCK, SOCIETY, AND CULTURE

As we have established, access to milk was important but uncertain in the old society. When it was felt to be threatened it was necessary to explain the misfortune so that one could find a remedy for it. Perhaps the herd owner asked himself if he could be the reason for the cows going dry? Maybe he wondered if he had neglected them in any way, or if their pasture or feed had been inadequate or unsuitable in some way. Perhaps people did seek these types of explanations, which today are considered 'rational'. However, it is certain that if these 'rational' explanations did not produce satisfactory answers, then people searched for explanations of an emotional kind. Had any of the well-known traditional rules for handling milk been ignored? Perhaps one had forgotten to place steel in the milk cask; or had someone with the evil eye gone by when the cows were milked; or had one failed to cover the container of milk with an apron on the way from the barn?

These attempts at explanations seem to us to be specifically different from those which are associated with herd mismanagement or poor pasturage. But this was not the case in the old society; rationality was then judged according to other criteria than it is today when the hard sciences play a much more decisive role in our perception of reality. In pre-industrial society, it could be just as rational and reasonable to adhere

196 *Milk and milk products*

to these older conceptions and rules, as to search for physical or medical explanations; the concepts of *belief* and *knowledge* had meanings other than they do today.

The next step in the attempt to explain the shortage of milk led one outside the barn, or one's own farm. Could some other person or power lie behind it all, perhaps? Was there someone in the neighbouring area who in one way or another went beyond the bounds of normality, someone who was rumoured to be different from the other people in his community? 'Yeah, there was something a bit strange about that woman who lived alone in her cottage far away from all respectable farms! How did she obtain her supply of milk? Could it be that my own milk luck has been transferred to her and her single skinny cow? But how could this have happened, what powers did that little crone possess, and how did she obtain them?'

In this way the search for an explanation—and also a scapegoat—could begin, and what followed is clearly to be gleaned from our archival notes and from the protocols of the witch hunts and trials.

I mentioned previously that the careful and jealous monitoring of each other in the old society was due to the fact that the society was built upon a delicate balance between resources and shortages, between luck and misfortune. When the balance was maintained life was 'normal', but if it was disturbed, then both the individual within the community, and ultimately society itself, were threatened. This idea was built upon what the American anthropologist George Foster calls 'the idea of limited good'. According to this theory, society has at its disposal a determined amount of resources and luck. If someone had a larger share of luck than that which he was expected to have according to the existing balance of resources, it was at the expense of others. Simply stated, luck was taken from one farm and transferred to another.

The balance between luck and misfortune was considered to be a sort of law of nature, or an order created by God. Power of a superhuman nature was required to break this order, and only the Devil could aid with such power, and for that reason arose the belief that the Evil One lay at the root of the problem.

This entire system of belief was very well anchored in pre-industrial society, and was concerned with much more than the stealing of milk from cows. It should rather be considered as an important part of the cosmology or perception of reality which characterised the old Swedish peasant society. But how widespread were these ideas? Who espoused them? How strong and real did they seem to people? There are no simple answers to these questions. Belief in what we today call the supernatural certainly varied socially and regionally, and it also changed over time. I also feel that the power of these ideas, that is to say, the reflected conviction, also changed from time to time in the mind of the individual. When things were as they should be, i.e. when the cows provided milk as expected, when the harvest was successful, and when one was as healthy as one deserved to be, that is to say, when the balance was maintained, then I do not believe that these ideas played an especially significant role in people's everyday life and consciousness. People certainly paid attention to certain 'irrational' rules of behaviour and followed certain rituals—when they remembered them—but I do not think that they necessarily firmly believed in what they did. Some of these rituals were probably even laughed at. It was just then when some misfortune struck that these ideas came into play with the same force as more observable causalities. In difficult situations a regression to perhaps primitive models of explanation occurred. People resorted to the safety-net which popular belief provided against threatening sudden changes. Perhaps this is how all folk beliefs function, regardless of whether or not they were, or are, part of the old or modern society? Then, as now, they functioned in conjunction with a struggle for security and cultural identity.

REFERENCES

G. Foster, 'Peasant Society and the Image of Limited Good', *American Anthropologist*, **67**: 2 1965.

O. Löfgren, 'The Reluctant Competitors: Fisherman's Luck in Two Swedish maritime Settings', *Maritime Anthropological Studies*, Vol. 2, no. 1, 1989.

Å. Nyman, 'Magisk tjuvmjölkning', I: *Atlas över svensk folkkultur: II. Sägen, tro och högtidssed* (eds. Å. Campbell and Å. Nyman), Uppsala, 1976.

G. Ränk, *Från mjölk till ost. Drag ur den äldre mjölkhushållningen i Sverige*, Stockholm, 1966.

J. Wall, *Tjuvmjölkande väsen I. Äldre nordisk tradition* Uppsala, 1977; *Tjuvmjölkande väsen II. Yngre nordisk tradition* Uppsala, 1978.

18 Milk and milk products in a woman's world

Ann Helene Bolstad Skjelbred

Due to natural conditions, which in large parts of the country were unfavourable for growing grain, pre-industrial farming in Norway laid great emphasis on animal husbandry for the production of food. For the most part, milk-producing animals such as cattle and goats were kept for the varied staple products from their milk. Sheep were kept for wool and meat, and both cattle and goats were important also for the supply of meat. To secure a sufficient supply of all these products was, therefore, of the utmost importance. The rich tradition surrounding the production and use of milk and milk products, and their place in women's world, is the subject of this paper. The analysis is based on folklore material, traditions documented from the middle of the last century until roughly the Second World War.

DIVISION OF LABOUR

In pre-industrial farming, the strictly gender-based division of labour gave women responsibility for the daily care of the milk-producing animals and the milk products. In large parts of Norway, natural conditions have dictated an extensive utilisation not only of the home fields, but also of the outfields for rough grazing of domestic animals. In fact less than three per cent of the total land area of Norway is arable land; mountain dairy farming has, therefore, during the stated period, been a vital part of animal husbandry in our country, as in other parts of Scandinavia. This meant that as soon as the snow was gone, and the grass had grown sufficiently for grazing, the domestic cattle were driven from the home farm to the mountains, where there was what was called a *saeter*, actually a mountain farm situated near water and grassy fields. One had to get permission from the 'invisible folk' to erect farm buildings there, and on arriving at the farm in summer, one had to ask their permission to stay there for the summer season. The rich mountain grass was used to fatten the animals and to produce as much milk, butter and cheese as possible for the coming year. There was also some haymaking at these mountain farms and the saved hay was usually brought home on sleighs during the winter.

During winter time, as many animals as the farm could possibly keep alive on fodder collected in the home fields and outfields, were stall-fed, and the rest were slaughtered. The limit for what was considered 'possible' could be stretched rather far, however. The term 'starve-feeding' implies that the domestic animals were kept alive on as little fodder as possible, with the result that when they were let out in springtime, they were so feeble and weak that their legs could break.

Women's role in the production of the important milk products is self evident, and procuring milk and milk products, therefore, constituted a vital part of women's daily life. Of the several important aspects connected with this activity, I shall here concentrate on knowledge and communication. Knowledge exceeds technical skill, and encompasses factors such as consciousness of the existence of supranormal power, the power of magic, and how to use it and deal with it. But knowledge is not enough; communication with supranormal powers, with people considered to have the power to destroy by magical means, and with women of the community, are also vital factors necessary to ensure a satisfactory personal result both economically and socially.

THE ELEMENT OF UNCERTAINTY IN DAIRY FARMING

Natural conditions dictated by climate and soil, set some vital limits for the yield from farming and animal husbandry in pre-industrial societies. Mountainous Norway with its northerly geographical position, its long winters and short summers, is on the edge, one might say, of where one can possibly achieve a reasonable result from farming and animal husbandry; dependency on nature, and strictly limited resources, constitute dominant traits in Norwegian folklore. Equally dominant is the belief that those limited resources had to be shared equally, in accordance with popular ideas of what was right and equitable, not only among the members of the human community, but among the invisible folk as well.

In addition to natural conditions, technical skills in animal husbandry and dairy farming also play important roles, and create distinctions between those who are industrious and those who are less so. Outstanding skills which gave better products than what was considered 'normal' were, however, not automatically welcomed.

The element of uncertainty involved in animal husbandry and dairy farming, allowed room for belief in luck and in disruptive forces in the guise of both the invisible people and fellow human beings. This meant that animal husbandry and dairy farming were professions where individual skill was not necessarily considered the dominant factor responsible for a given result. The importance of co-operation with the invisible forces, and of staving off any dangers from nature in the form of wild animals,

200 Milk and milk products

and from envious, and therefore, evil people, were also vital factors in securing a good result. In their profession, women had to consider all these factors, and their skill was, therefore, not only a technical one, but consisted perhaps more in having the ability to manoeuvre between what was considered a 'normal' outcome to their labours, and the ambition to prove themselves within their given field of responsibility.

It is important to emphasise that the dominant belief system was, that the world consisted of supranormal forces, and that in daily life everything was dependent on coexistence with these forces. In relation to the successful production of farm products from domestic animals, it was essentially the women's responsibility to establish and conduct satisfactory co-operation, both with supranormal beings and with people with supranormal powers. This balance was achieved partly by the use of magic. To know and observe the ways and means of promoting and securing ample and satisfactory milk products, and of averting constantly impending dangers, was, for the most part, women's responsibility. Accordingly, the folklore corpus concerning the production of milk and milk products consists mostly of women's folklore.

WOMEN AND KNOWLEDGE OF MAGIC

From Valdres, a valley in the southern part of Norway, the following has been documented: 'Most of the older dairymaids used to sing while they were milking. They were sure the cows liked this song, and it was important to make them comfortable so that they would give a lot of milk and not kick the bucket. It was believed that the cows could hold the milk back if they did not feel comfortable. This was one of the songs the dairymaids could sing: "Are you my sweet one, you? Are you my darling, you? You are the best one to give in my bucket, you!"'[1]

Thus making the cows feel at ease was considered one way of securing a satisfactory result and of getting a lot of milk. Another way was to spill some milk on the floor for the invisible people. One could also milk in a sunwise direction, milk on a piece of steel, or on an inherited ring placed in the bucket, or on a stone said to be made by the dwarfs fitted into the bottom of the bucket, or on a cross engraved on the bucket to ensure that the milk would not be ruined. To make the sign of the cross after one had milked was another protective measure which could be taken, like saying the name of Jesus Christ over the milk as one finished. To avert bad results it was forbidden to milk into a dry bucket, or to milk after sunset, or to carry the bucket of milk over running water. When the milk was brought into the house, it had to be covered with a piece of cloth.

Some of these ways and means of promoting and securing the milk supply must be viewed in relation to the invisible people. To share some of the milk with them was considered a good measure to keep the supranormal powers friendly. To share the cow-house and the dairy with them was also considered important if one wanted to have good luck with the dairying. All in all, therefore, the relationship with the invisible people was one marked by co-operation; a good relationship with them would give good results with the domestic animals and with the dairying, while provocative or unfriendly actions towards them would bring bad luck and even disaster. Folklore material commenting on the relationship between humans and the invisible people mostly depicts co-operation between women: the dairymaid or the farmer's wife in co-operation with the fairy woman.

Other ways and means mentioned in folklore must be viewed in relation to the other impending danger: envious people with evil powers, and the will to use their powers to adversely affect others, and to take more than their own share of the common but limited goods. To have luck with the domestic animals meant to get a lot of milk and butter, and when one achieved this, one was said to have 'milk-luck' and 'butter-luck'. By means of magic one could secure one's own milk—and butter-luck, but people with evil powers could, according to folk belief and legends, steal both one's milk—and butter-luck in order to enrich their own supply. Some of the legends depicting the evil of witches, have as their main motif the theft of milk and menace to domestic animals. In the legend of the daughter of the witch, the witch's daughter milks the neighbour's cow until it falls down dead (Migratory Legend 3035),[2] and in the legends of the witch making butter, the witch steals cream from others to make butter for herself (Migratory Legend 3040).[3] This fear of theft by evil people is documented in both beliefs, memorates and legends of the witch's rabbit or cat. It was believed that evil people by means of magic, could make themselves what was called a milk-rabbit to send out and suck milk from other people's cows and bring it home to its owner.[4]

To have 'butter-luck' meant not only to get a lot of butter with a delicate yellow colour, but also to have luck with churning. Due to the variable quality of milk especially in the past, churning could be very hard work indeed, and if butter did not form this situation could be considered the result of someone's evil and magical powers, and it had to be dealt with by means of counter-magic. This could involve regular magical prohibitive acts such as making the sign of the cross, or placing a piece of silver in the bottom of the churn. Another form of prohibitive magic was to find the witch responsible for the loss of 'butter-luck' and thus deprive her of her evil powers. Although people with evil powers were of both sexes, it is a dominant trait in ideas about animal husbandry, and especially about the production of milk and milk products, that harm is perpetrated by women against women. The relationship is, in this case, not one of co-operation as with the fairy people, but rather a question of power; counter-magic had to be pitted against the perceived magical acts, and the women had to be strong enough to stave off the evil which could be inflicted by people with evil powers.

WOMEN AND KNOWLEDGE OF FOLK MEDICINE

Knowledge of folk medicine is another area of female dominance. The remedies used for healing both man and beast show that people used ingredients that were ready at hand. Among the different components used to heal superficial as well as serious ailments, one also finds food, e.g. milk from cows and goats; cream, both sweet and sour; butter; different types of cheese; porridge made of sour cream; or the fat dissolved from, and floating on, sour cream porridge. Butter was also used against snake bites and sea sickness, for example, while cheese and butter were good remedies against stomach-ache. Sour cream could be used to protect against blood poisoning, and the fat from the sour cream porridge was good for stomach complaints in babies.

It was probably, also, mostly women's responsibility to produce and have in store, the right remedies for ailments in the milch cattle. Different ailments of the udder, for example, hindered milking, or caused the milk to deteriorate, and had to be healed. Remedies for this included steam from boiled hayseed, or steam created by dumping a cold stone into boiling cow's urine. Putting butter or grease on the udder was also considered an effective remedy. While it is difficult to detect a pattern in the use of milk and milk products in folk medicine the conclusion must be, however, that such products were always at hand for women to use when something was needed to cure different ailments in the cow-shed or in the house.

WOMEN, WORK AND SOCIAL EVALUATION

Folklore makes clear that to secure a satisfactory, product, both in terms of quantity and quality, more than just technical skills were needed. A lack in quantity meant a shortage of food for the household during the long, non-productive period of the year, and showed that the women had failed in their responsibilities in this respect. It showed a deficiency in expected competence, but in view of the emphasis placed by tradition upon supranormal reasons for such shortcomings, human incompetence was played down, and it was the relationship with the fairies, and the ability to stave off misfortune brought upon the society, both by the fairies and by people with supranormal powers, which were considered to have been faulty.

But there were occasions when women had special opportunities to demonstrate their competence regarding milk and milk products. The custom of bringing food to the neighbours at events like births, weddings and funerals was strictly organised in pre-industrial times. The community was organised so that every farm was represented in the limited groups responsible for furnishing food for these large and important events in life; the organised groups were smaller for births, and larger for weddings and funerals. This custom gave women a chance to show the best of their products in public. At birth the fat sour cream porridge was customarily brought to the woman in childbed. The porridge was said to butter the inside of the stomach, something that was considered very good. But clearly the occasion was also an opportunity to compare the porridge from different households. It was a sign of rich cream if the porridge set free a lot of fat. To produce a rich cream, and to be able to make a proper sour cream porridge with the flat floating on top, is still, even today, considered evidence of being a very skilled housewife.

Sour cream porridge was a dish used only on special occasions, and some varieties of butter and cheese were also only so used, such as those which were part of the 'sending' to weddings and funerals. As the dishes were brought into the house they were placed on the table for everyone to see and comment upon. Whitish butter gave the giver but little credit, while the best and nicest looking butter achieved the place of honour on the table; 'In the basket brought to the wedding the butter was an obvious ingredient, and for supper the balls of butter in their bowls were placed on the table. The biggest and most distinguished was placed at the head of the table by the bride and groom. It was a great honour for the owner of this butter to achieve such an honourable position'.[5]

Not much imagination is needed to understand how such events could, and did, rank the housewives of the community according to their perceived competence, or lack of it, in relation to dairy products.

Other events at which women could demonstrate their efficiency in regard to milk and milk products, were what were called 'mountain farm parties', which took place during the summer. The dairymaids staying at the mountain farms met and the hostess for the day displayed her products.[6] The importance of this demonstration can be gleaned from the following information from the Valdres valley:

> About 60 years ago there lived a woman in this valley who wanted to be the best in both this and that. Once when she held a dairymaid party at her mountain dairy farm she really wanted to show off. The party was lavish, and after they had eaten, they all went to view her milk products. There were cheeses in rows and rows. But the tubs filled with butter were the most surprising. They all knew that she had only a couple of cows! One of the dairymaids was brazen enough to lift one of the tubs and it was as light as a sack of feathers. She then put her finger along the rim of the tub and it turned out that there was only a thin layer of butter on top of a pile of linen shreds. This incident gave opportunity for some laughter and a lot of talk afterwards.[7]

At other mountain farm parties people, especially young people, from the community came to visit, to taste the fresh dairy products and to dance and have fun. This was another chance to demonstrate skill in cooking sour cream porridge, and to display both the quantity and the quality of the milk products ready to be brought back to the main farm.

204   Milk and milk products

The dairymaids prepared in a special way for going home when the summer season in the mountains was over. They produced a large quantity of what was called 'homecoming cheese', or gift cheese or something like that. These were small pieces of cheese, not more than about a mouthful in size, and often resembled small figures like stars, animals or the like. Some of them were indeed little works of art! On their way home with the herd and the milk and milk products, they handed out these small pieces of cheese to people they met, or at the farms they passed on the way. The handing out of the cheese confirmed that the end of a productive season had come to an end, a season involving mainly women's responsibility, when women were also given the opportunity to demonstrate their skill and competence in many ways.

CONCLUSION

One can look upon the field of animal husbandry as an arena in which women were obliged to perform clearly defined roles as a consequence of the general division of labour. In this arena their performance was judged according to the quantity and quality of the products they produced, and of these, milk and milk products were not the least important for the survival of the whole household. But the quantity and quality of the products were dependent not only on the women's human skills, but also on their competence in magic in order to secure a balanced working relationship with the invisible people, and to stave off any threats from humans with potentially harmful magical powers. The result was also dependent upon women's competence in the field of folk medicine. Women's competence, therefore, had to be many-sided in order to do justice to the proverb, 'Prosperity lies in the housewife's hand'.

There is no reason to believe that the need or desire to win acclaim for one's efforts was less in former times than it is now. These requirements could be satisfied not only by means of a well-filled food storehouse, but also through direct confrontation and competition with other women at gatherings and parties. When one's butter or sour cream porridge was highly esteemed, one achieved high social status in the community of women on several levels; when it was not, only shame and low esteem were the women's lot.

REFERENCES

1. K. Hermundstad, *Ættararv. Gamal Valdres-kultur*, IV, Oslo, 1950, (Norsk folkeminnelags skrifter 65), 27–28.
2. See R. Th. Christiansen, *The Migratory Legends. A Proposed List of Types with a Systematic Catalogue of the Norwegian Variants*, Helsinki, 1958. (FF Communications N:o 175).
3. *Idem.*

4 S. Solheim, *Norsk sætertradisjon*, (Oslo 1952), (=Instituttet for sammenlignende kulturforskning. Ser. B. Skrifter 47), 311–313. See also J.I. Wall, *Tjuvmjölkande väsen. I. Äldre nordisk tradition*, Uppsala, 1967. (Acta Universitatis Upsaliensia. Studia Ethnologica Upsaliensia 3)
5 K. Djupedal, *'Dei va best, dei so va omtrent i balanse'. En undersøkelse av verdensbildet i Selje i 1920*. Magistergradsavhandling, Etno-folkloristisk institutt Bergen 1985, 132.
6 Solheim *op. cit.*, 520–521.
7 *Ibid.* 521.

SELECT AND ANNOTATED BIBLIOGRAPHY

R. Asker, 'Yste ost og kinne smør', *Drammen turistforenings årbok* 1965, 24–39. III. (On daily life and work at the mountain dairy farms; tools and products)

T. Austrumdal. 'Um stølar og stølsbruk i Dalarne i eldre og nyare tid', *Årbok for Rogaland historielag* 27, 1941–42, 30–50. (Mountain dairy farms and mountain dairy farming in Dalarne, in the south-western part of Norway).

A. Bakken, 'Ysteriet i Ramnes. Grunnlagt 24. februar 1860', *Vestfoldminne* 1960, 38–41, III. (About a cheese factory, established in 1860 in Ramnes, a small region to the west of the Oslo fjord).

J. S. Bang, 'Om Suurmelks Afbenyttelse til den saakaldte norske Gammelost og Syre', *Nordisk landvaesens og Landhuusholdnings Magasin* Bd. 4, København 1802, 169–175. (On the use of sour milk for making 'gammelost', a fully matured, highly pungent cheese and whey).

H. Drabløs, 'Melkehandelen i Ålesund frem til 1940', *Skrift. Aalesunds museum* 14, 1984, 29–34. III. (On the sale of milk in Ålesund, a city on the West coast of Norway).

O. Evenstad, 'Beskrivelse over (den efter Ostlavning overblevne) Valle, paa norsk Myse, og dens Indkogning til Myssmør-Ost og (den efter suur Melk overblevne) Valles Anvendelse til Drikke', *Det Kongelige Danske Landhuusholdnings-Selskabs Skrifter*. Nye Samling. Bd.2, København, 1811, 197–122. (The author gives a description of how to make cheese from whey, or how the sour whey can be used for drinking.)

S. Funder, 'Om gammelost', *Naturen* 72, 1948, 165–171. (On matured, pungent cheese made from sour whey called 'gammelost').

S. Funder, *The Chief Molds in Gammelost and the Part Played by Them in the Ripening Process*, Oslo, 1964.

J. Grude, *Stølsdriften på Vestlandet*, Stavanger, 1891. (A historical description of mountain dairy farming: buildings and architecture, tools, and different forms of churning and cheese-making).

F. Grøn, *Om kostholdet i Norge indtil aar 1500*, Oslo, 1927. (=Skrifter utg. av Det Norske Videnskaps-Akademi i Oslo, 1926.11. No. 5). (A broad survey of food in Norway up to 1500. Ch. 4 deals with animal husbandry and products from domestic animals; milk and milk products, 89–104).

F. Grøn, *Om kostholdet i Norge fra omkring 1500-tallet og opp til vår tid*, Oslo, 1942. (=Skrifter utg. av Det Norske Videnskaps-Akademi i Oslo, 1941.11.No. 4). (A broad survey of food in Norway from 1500 to the present time; milk and milk products, 82–101.)

J.S. Gunnerus, 'Oeconomisk Afhandling om alle de Maader, hvorpaa melken nyttes i Norge', *Det Kongelige Norske Videnskabers Selskabs Skrifter*. Femte Deel, Kiøbenhavn, 1774, 26–152. III. (An economic thesis on different uses of milk in Norway. The material is mainly from Trøndelag and is partly based on the author's own knowledge and experience. The work is divided into three parts: (1) A description of different churns and other dairying objects and how they are used, followed by a description of the necessary hygiene and further advice on how to obtain a satisfactory result, and scorning ideas considered superstitious, among the common people; (2) A description of 19 different kinds of cheese; and (3) 30 different milk dishes).

O.C. Holch, 'Planternes Indflydelse paa Melkens, Smørrets og Ostens Beskaffenhed kortelig forestillet', *Det Kongelige Norske Videnskabers Selskabs Skrifter*. Femte Deel Kiøbenhavn 1774, 525–538. (The different effects on milk, butter and cheese, in relation to richness, colour, and taste. of different plants. The author is scornful of a series of described superstitions).

H. Hyldbakk, 'Sæterbruket på Nordmøre', *Årsskrift for Nordmør historielag*, 1941, 5–20. (Mountain dairy farming in Nordmøre, a western region of Norway. Daily life, stories, legends, and beliefs).

P. Jæger-Leirvik, '"Tettmelk" og "Settmelk"', *By og Bygd*, 1959, 43–50. (On milk thickened by using butterwurt (*Pinguicula vulgaris*)).

R. Marcussen Kielland, *Milk shake og gamalost: melkeprodukter, tradisjoner, oppskrifter*, Oslo, 1988. Ills. Figs. (Traditional use of milk, but mostly on contemporary use of milk and milk products in Norway).

R. Marcussen Kielland, *Takk for melken. Om utvikling og bruk av meieriprodukter*, Oslo 1976). Ills. Tab. (On the historical and contemporary development and use of dairy products).

J. Kleiven, 'Saeterbrunne i Vestfold', *Vestfoldminne* **4**, 1941–42, 97–143. (Mountain dairy farming in Vestfold, a district west of the Oslo fjord).

H. Magerøy, 'Fraa mjølkehandelen si soga', *Nordmøre historielags årbok* 1931, 77–82. (On the history of the sale of milk).

L. E. Myraker, *Gudbrandsdalsost og Gudbrandsdalsysterier. En oversikt over utviklingen fra 1860-årene og fram til vår tid*, Gjøvik, 1945. (Industrial production and sale of cheese from Gudbrandsdalen from different cheese factories in the valley).

R. Pedersen, 'Pultosten—et klenodium fra Hedmarksbygdene', *Lautin* **16** 1990, 211–215. III. (On a traditional soft, sharp cheese spread made from sour whey).

L. Reinton, *Sæterbruk i Noreg*. I-III. Oslo 1955–1961. (=*Instituttet for sammenlignende kulturforskning. Ser. B. Skrifter 48:1–3*) (An extensive survey of mountain dairy farming in Norway; history, vernacular architecture, tools, the mountain dairy farm's importance for production of milk and milk products, and harvesting of other means of subsistence).

P. Smørvik, *Melkstell*, (Mosjøen 1979. III. (On traditional tools and production of milk products).

S. Solheim, *Norsk sætertradisjon* Oslo 1952. (=Instituttet for sammenlignende kulturforskning. Ser. B. Skrifter 47) (An extensive survey of folk traditions: beliefs, rites, legends, connected to pre-industrial mountain dairy farming).

S. Solheim, 'Stølssengner', *Stavanger turistforenings årbok* 1943, 29–42. (Legends about mountain dairy farms).

O. Sopp, *Undersøgelser over ost og ostegjæring*. I. Indledning. Surmelkostene, Kristiania 1905, Ills. (A broad survey ranging from home-production to cheese factory production of 'gammelost' and other cheese from sour milk).

T. Steinbakke, 'Frå gåmålostysting til dampysteri', *Årbok for Gudbrandsdalen*, 1961, 121–123. (From home production of mature, pungent cheese made of sour whey to the first steam-powered cheese factory in Skjåk, 1896).

I. A. Streitlien, 'Seterbruket', *Tynset bygdebok*, Vol. 3, Tynset, 1978, 315–372; III. (A broad-ranging survey of mountain dairy farming in the Tynset region in south-eastern Norway towards the Swedish border, from historical sources dating from 1661 to the 1970s. A description of buildings and vernacular architecture, daily life, and the production of milk, cheese and butter).

J. Sundt, 'Da gardsseparatoren fikk innpass i Gudbrandsdalen', Årbok for Gudbrandsdalen **54**, 1986, 52–59, Ills, (The introduction of the cream separator in Gudbrandsdalen in the 1890s).

J. Sundt, 'Den første omsetning av gudbrandsdalsost', Årbok for Gudbrandsdalen, **58**, 1990, 89–100, Ills. (The introduction of the newly-developed *gudbrandsdalsost*, a brown cheese made from sweet full-cream goat and cow's milk, into the market in the 1890s).

J. Sundt, 'Feitostproduksjon i Gudbrandsdalen', Årbok for Gubrandsdalen, **59**, 1991, 77–87, Ills. (The industrial production of full-cream cheese made from cow's milk in Gudbrandsdalen).

K. Sydnes, 'Stølar og stølsliv i Aakrefjorden', *Sunnhordland. Tidsskrift*. **19**, 1938, 3–41. (Description of daily life on different mountain farms in Åkrefjorden on the west coast of Norway, incorporating stories and legends).

N. Tveit, 'Mjølkebruk og mjølkesamskipnad i Noreg', *Norsk Aarbok* **15**, 1934, 33–42. (Dairy farms and dairy organisation in Norway.).

J. Veka, 'Stølsliv i Ryfylke før og no', *Stavanger turistforenings årbok* 1942, 36–51. (Mountain dairy-farming in Ryfylke, a district in the south-western part of Norway).

E. Wekre, *Melkeomsetningen i Oslo gjennom 75 år, 1872–1947*, Oslo, 1948. (On the sale of milk in Oslo over 75 years, 1872–1947).

19 Women, milk and magic at the Boundary Festival of May

Patricia Lysaght

> And he made peace for the sake of his cows and his people (*do righne síth dar ceann a bhó 7 a mhuinntire*).
> (*Annals of Connacht for the year 1125 A. D.*)[1]

I

This is one of a number of quotations cited by A. T. Lucas in his seminal work *Cattle in Ancient Ireland* (1989), to demonstrate the importance of cows in early and medieval Irish society.[2] The preoccupation with cows and their well-being is evident in the works of Irish chroniclers, hagiographers, historians, poets and other writers, and the annals, law tracts, lives of saints, historical narratives, praise poems, tales and anecdotes in prose and verse, teem with allusions to cows as yielders of milk.[3] The cow also served as a standard and measure of value, and the number of cows in a person's possession was an indicator of his poverty or wealth. Raiding far and near, and friend and foe alike, for cows, seems to have been almost a way of life from the dawn of Irish history to early modern times, and indeed cow-raiding is the theme of the greatest of the Irish epic tales, the *Táin Bó Cuailnge* or 'Cattle Raid of Cooley', the oldest surviving manuscript version of which dates to about the beginning of the twelfth century.[4] *Cáin Lánamna* or 'The Law of Couples', a special law tract on marriage dating from about the eighth century, lays down that a husband may not alienate cows without his wife's consent,[5] and the stipulations governing the division of dairy products in divorce cases reflect a greater contribution on the part of the wife in their production.[6] A variety of milk products are mentioned in the ancient laws of Ireland,[7] while milk and a wide range of dairy products are prominent among the many kinds of food mentioned in the early-medieval Irish *Aislinge Meic Conglinne* or 'The Vision of MacConglinne', a satiric text rich in food imagery, and probably the most important source of information about food in medieval Ireland.[8]

It is not just the indigenous literature, however, which testifies to the value placed on milch cows as a source of food by the Irish people down the centuries. Various English commentators on Ireland in the sixteenth and seventeenth centuries, emphasised the importance of milk and milk products in the diet of the people. Among them was the poet Edmund Spenser, who, as secretary to the lord deputy during the Tudor conquest of Ireland in the sixteenth century, was granted considerable areas of sequestered land in the province of Munster. He stressed the necessity of waging war against the Irish rebels during the non milk-producing periods of winter and early spring, or of distraining, harrassing, or destroying the cow-herd in order to force the rebels' submission for want of food during the following summer season.[9] In the early decades of the seventeenth century, the English historian Fynes Moryson, commenting on the Irish predilection for a variety of milk preparations collectively called *bánbhia*, literally 'whitefoods', but anglicised 'whitemeats', stated that they carefully attended to their cows and fought for them as for religion and life, and even when they were almost starved, they would only slaughter a cow that was old and did not yield milk.[10]

But milk products were not just primary elements of the Irish diet in the time of Spenser and Moryson; butter was already an item of international commerce in the mid-sixteenth century. To what extent that was so is uncertain, however, but it was clearly of significant quantity since its export was prohibited in the 1550s, and severe export duties imposed on it in the 1560s and 1570s by the English crown.[11] In the early seventeenth century, export duties were imposed by the crown on consignments of butter destined for countries other than England.[12] Butter exports from Ireland increased substantially in the seventeenth century[13], and attempts to introduce legislation to regulate the butter trade were made around the middle of that century[14]—by that time the art of preserving butter by barrelling was known in Ireland.[15] But it was in 1698 that the first important butter Act was passed, the preamble of which stated, that butter was one of the principal commodities produced in Ireland 'not only at home but very great quantities thereof are transported beyond the seas.'[16] This, and a series of Acts in the course of the eighteenth century, were concerned *inter alia* with the size, weight, and capacity of butter casks, imposing penalties for excessive salting of butter, and directing that the wood used for butter casks should be either seasoned ash or oak.[17]

The extent of the Irish export trade in butter, and the variety of destinations—apart from Britain—that it reached in Europe, Africa, the 'Atlantic Islands' and the Americas, in the eighteenth century is truly astonishing.[18] At that time—and indeed for much of the following century—Irish butter had pre-eminent position in the international provisions market because of its good keeping quality on long sea vogages, and there was 'a demand and a market for it in almost every country where British trade penetrated'.[19] In the middle of the following century the annual export of Irish butter was estimated at two million firkins[20] of seventy lbs. weight each, with a money value of about five million pounds.[21]

In the eighteenth century, Cork city and port handled almost half of the total amount of butter exported from Ireland.[22] In the nineteenth century, although other large butter markets existed in Limerick, Tipperary, Waterford and Belfast, Cork city, with its extensive hinterland of the rich pastures of the Golden Vale, a guaranteed annual production of large quantities of butter, as well as a suitable port, continued to dominate the provisions export market. The remarkable success of the Cork Butter Market for upwards of a century, stemmed from the quality controls set in place by a voluntary organisation of merchants, The Committee of Merchants, founded in 1769, which resolved: 'That we shall ship no butter which shall not be publicly inspected, marked and branded'.[23] The basis of the system adopted in Cork was the classifying of all butter into qualities, three in the early years and six during the later years. About 1830 the symbols used to denote quality were as follows:

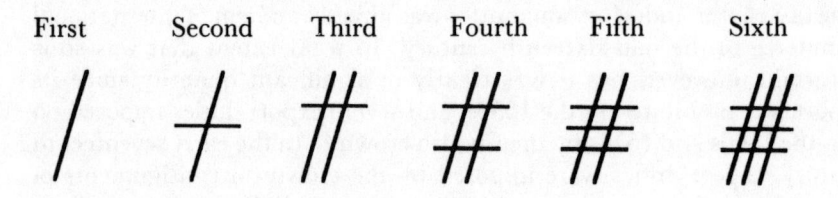

First	Second	Third	Fourth	Fifth	Sixth
/	+	#	#	#	#

(Source: W. O'Sullivan, *The Economic History of Cork*, (Cork 1937), 275.)

As a result of this quality control system, which extended also to the quality of butter casks and packing methods, Cork had premier rank in the Irish butter market, and by the middle of the nineteenth century almost two-thirds of the butter exported from Ireland was handled by the Cork market. For over a hundred years from its inception, the internal control and management of the Cork Butter Market was vested essentially in the voluntary organisation, the aforesaid Committee of Merchants. It was not until 1884 that the operations of the Market were formally regulated by statute when an Act was passed 'for the permanent establishment, regulation and management of the Butter Market of the City of Cork'.[24] By that time the Market itself had fallen victim to quality control abuses, as well as antiquated methods, and was no longer a potent force in the butter trade, though Cork city remained a centre of the export trade in provisions.

The demise of the Cork Butter Market was hastened by the rise of the creamery system and the change to mechanical dairying, which may be said to date from 1885 in Ireland, with the introduction of the continuous centrifugal separator invented by the Swede, Gustav de Laval.[25] A number of creameries were established in the Golden Vale area in the mid-1880s. These were set up on joint stock principles where farmers and others held shares and received a dividend.[26] The first co-operative creamery was established at Drumcollogher, Co. Limerick in 1889,[27] and others soon followed. Today 97 per cent of the Irish liquid milk industry is controlled by co-operatives.[28]

Dairy cows and milk production remain of paramount importance in the Irish economy to the present time. In 1989 the production of milk constituted 35 per cent of total agricultural production in the Republic of Ireland[29]—almost double the EC average. For the same period butter production accounted for 59 per cent of total utilisation of whole milk (against an EC average of 33 per cent), but cheese production at just 13 per cent was only half the EC average.[30] Milk and milk products also remain important in the Irish diet, although in Ireland, as in the other countries of the EC, the consumption of such products is declining. Nevertheless, the Irish remain by far the highest consumers of liquid milk in the EC: in 1990 the consumption per capita in the Republic of Ireland of whole milk was 166.9 kg.—more than treble the EC average consumption,[31] while skimmed and buttermilk consumption was 15.7 kg. per capita,[32] which was double the EC average at that time. However, there has been a drastic drop in butter consumption in recent times as Irish people have become aware of the need for healthier eating patterns, and there has also been a corresponding rise in the consumption of margarine and butter spreads: in 1990 butter consumption at 3.5 kg. per capita was less than one third of the 1984 peak consumption rate of 12.6 kg. per capita, and also less than the EC average of 4.7 kg.[33] Cheese consumption, traditionally very low in Ireland, is slowly increasing; still, the 1990 consumption rate of 5.5 kg. per capita is only about one third of the EC average.[34]

II

So far, evidence has been adduced to show the importance of milk and milk products in the early, medieval and modern Irish diet and economy. For the early period this consists mainly of textual references to the predominance of cattle, particularly milch cows, in the economy, and to a wide variety of milk preparations and products. The testimony of foreign observers of the Irish scene provide valuable evidence for the medieval period, and recent historical works and official data help to fill in the picture for the late medieval and modern periods. But the importance of cows, milk and milk products in the diet and economy of the Irish people from ancient to modern times may also be assessed in another way—in mythological terms. There is no shortage of source material to support such an assessment, though of course the data, which becomes more apparent from the sixteenth century, really becomes abundant from the nineteenth century onwards, as we shall now see.

Bealtaine: the boundary festival

Indicative also of the importance of milk and milk products in the ancient and medieval Irish economy and diet, are the numerous literary references to the sensitivity of cows to mystical and occult forces, benign or malign,[35] an idea which has, with some variations, persisted into modern times. From about the twelfth century we get glimpses of a belief system—not unfamiliar to a twentieth century Irish country person—concerned *inter alia* with the susceptibility of milch cows to good and evil influences, particularly at *Bealtaine*, the boundary festival of May, which heralds the beginning of summer, and thus also the commencement of the milking season in Ireland.[36] For the duration of May eve (*Oíche Bhealtaine*) and May day (*Lá Bealtaine*)—the period of transition between the winter and summer seasons of the year—one's dairying luck was felt to be very precariously balanced, and could be either lost, maintained or promoted during this ambiguous time. That fears in relation to the loss of milk and butter luck predominated at this time is evident from the rich corpus of beliefs, customs, rituals and legends concerning milk and milk-'profit' theft, and in particular, the measures taken to protect one's milk and butter luck, or to recover one's stolen 'profit', as the case may be.[37]

The main repository of relevant primary source material, which is both quantitatively and qualitatively suitable for a full-scale study of Irish attitudes to milk and dairy produce at the boundary festival of May for at least a century, is the archive of the Department of Irish Folklore, University College Dublin. Most of the material in relation to *Bealtaine* in the archive, including a questionnaire on the topic issued in 1947, has not yet been fully analysed or published.[38] This article is based mainly on that questionnaire material, and on the author's fieldwork in relation to the festival of May undertaken in recent years in the Irish countryside.

It is immediately obvious from an analysis of this material that in Ireland in the nineteen forties, some people still firmly believed that their milk 'profit' could be stolen during the festival of May. A correspondent from Co. Laois, in the midlands of Ireland, stated that although belief in milk and butter stealing had almost died out in the area, nevertheless, for some it was still a firmly held belief, and she adds:

> Miss Moran, a poultry and dairy instructess for Co. Laois for nearly forty years, told me she had been consulted on several occasions about getting back butter that had been *taken* . . .[39]

It is also evident that the milk stealing activities, and the vast majority of the protective measures taken to guard against 'profit' theft (both usually referred to as *piseoga*), were regarded as acts of magic, but not, it seems, involving a diabolical pact. The 'profit' thief was almost invariably believed to be a woman—a farmer's wife, or an independent woman farmer seeking to increase her own dairy produce at the expense of that of her neighbour's, and usually a neighbour or even a relative.[40] Protective rites began on May eve when the festival of May properly began, and continued until noon on May day, while rites connected with taking milk or milk profit were usually performed around sunrise on May morning—though some might take place between May eve and noon on May day. Both sets of rites could be performed either in the wild landscape—at particularly magical spots such as at the confluence of three rivers, or at the meeting point of three townlands or farms, or in the domesticated landscape—on the family farm, at wells, in the farmyard, outhouses or in the dwelling-house, though of course some rites might be performed in both landscapes. We shall now consider aspects of both 'profit' stealing and the protective rites in more detail.

Stealing the milk 'profit'

Measures to protect one's milk luck were taken throughout the milking season, and indeed throughout the year: on any occasion when milk was given away, a pinch of salt was usually added to it, and salt was also sprinkled on butter destined for the market before it left the farm. But it is evident from the source material that it was undoubtedly during the festival of May that peoples' fears were greatest in relation to interference, not just with milk quantity, but especially with milk quality, and the usual protective measures were intensified and special ones introduced at this time. Occasionally clearly expressed, but more often implied in the numerous references to the loss of *sochar an bhainne* or 'milk profit', or in those reports which state that the butter had been 'taken', is, that the milk quality i.e. the butter fat or cream content, has been meddled with. This is evident in those accounts which tell of churnings producing only froth and a foul or sour-tasting liquid.

Interference with milk quality is also emphasised in narratives detailing unproductive or difficult churning situations believed to have resulted from the May morning theft of substances symbolic of cream (often referred to as the 'top of the milk'), such as the dew on the grass, the surface water of the well or 'the top of the well', or water from a stream or river constituting a territorial boundary of some sort. Dew could be collected by a would-be butter thief in a number of ways. Simply walking across other peoples' land was considered sufficient as the thief involved could be walking off the dew and hence the milk luck.[41] Consequently, people all over Ireland intensely disliked seeing others, especially strangers and women, walking through their land on May Eve or May Day. A Co. Galway correspondent who wrote in 1947 that 'people who went into other peoples' land on May Day were still thought to be up

214   *Milk and milk products*

to no good ... to be trying to work evil magic', succinctly sums up the general reaction typical of the archival material.[42] Indeed, around the same period a Co. Cork farmer was quite unequivocal in his response to a query about this matter. The questionnaire correspondent who spoke to him wrote as follows:

> Some people will not allow any person to walk their land for any purpose on May Day. A man living about four miles from my own village of Knocknagree makes no secret of it that he would shoot any man or woman he'd catch walking through his land on May morning. He says 'what would bring them there, don't they know it is May Day, and don't they know what is said about such things on May Day? If they have no bad intention they won't be found doing it, and if they come with a bad intention don't they deserve to be shot? And the man that would shoot such a person would be doing a good act for the neighbours.[43]

Generally it is an object intimately associated with milk production that is used to gather the dew. What is most commonly employed is the cow-spancel or *buarach* (< *bó* 'cow', *árach* 'fetter'), usually a short piece of rope used to bind the cow's hind legs together during milking in order to immobilize her. Traditionally in Ireland milking has been regarded as a woman's job, and the spancel the *vade mecum* of the milker. Thus in view of the close sympathetic association between the woman and the spancel, and also between the cow and the spancel, especially since it was often made of a short two-ply rope twisted from the long tail hairs of cows, its use as a medium for milk-stealing is understandable.[44] Where a rope is employed to gather dew, the implication is that it is a spancel which is being used, and the following Co. Fermanagh account gives a fairly typical description of the May morning dew-gathering activity:

> A [man] on May morning, saw two women (neighbours) in his field dragging a rope across the field. It was very early, so he went to them and he recognised them as two women who lived alone a few miles away and who were infamous for these sort of activities.[45]

Further evidence of the vigour of the belief that the May morning dew on a neighbour's land represented his milk profit, and that this profit could be magically stolen from him through the appropriation of the dew by dragging, for example, a spancel, a rope, a cloth, or even a briar as we shall see below, along the dew-laden grass, is the following version of a legend common in Ireland. This version from south-west Donegal, tells of a priest going on a sick call who accidentally becomes a party to a magical milk-stealing rite being performed by a woman, who is 'gathering the dew' with a cloth. The legend also underlines the probable butter loss which any farmer in these circumstances might sustain by emphasizing the huge butter gains to the 'thief':

Women, milk and magic at the Boundary Festival of May   215

> ... There was a priest in Ardara long ago, and early one May Day morning he was going on a sick call out west to Luachros point. He saw an old Protestant woman drawing a piece of cloth after her through the dew and calling out that all the milk belonging to certain people should come to her. He called out that two-thirds of it were to come to him.
>
> Well and good. He went on and attended the sick person and when he returned home, his barn was filled with butter and milk. Next Sunday he announced from the altar that anyone who had lost his butter or milk should come to him and that it was all to be got in his barn. They came, and so he got all the butter and milk back to the people of Luachros point who had lost it.[46]

Indeed a farmer could put his own milk and butter supply at risk on May morning as the following legend from Co. Westmeath warns. In this narrative, the woman of the house is shown to have accidentally collected the May morning dew of the farm, and thus the farm's milk profit, by inadvertently trailing a briar along the grass causing it to become a medium for milk theft, which was then appropriated by an outsider thus putting her milk and butter-supply for the season at risk:

> A woman milked her cows in a field and was bringing the milk home along a road where a man happened to be fencing [i.e. constructing a fence]. A long briar which got attached to the lady's skirt trailed after her. As she passed, the man noticed the briar and cut it with the spade. The milk at once poured out over the the pail tops and was lost on the road. The woman at once threw herself on her knees, cursed the man and threatened to steep a sheaf for him.[47]

For a traditional farming audience the woman's immediate and powerful reaction portrayed in the narrative—involving cursing and a threat of black magic[48]—confirms belief in, and a fear of, milk stealing.

Other methods of milk stealing are also mentioned in the sources. Spancels might be used not only for gathering dew, but might also be placed in boundary water for the purposes of stealing the milk and butter of the cow herd which drank there. Women seen at boundary water on May morning were believed to be 'gathering butter', and the following legend from Co. Kerry tells how the *buarach* might be used in boundary water for that purpose. The collector reports:

> One woman told me (she is about 81 years old) that her father drove the cows to the water of three boundaries one May Morning. The cows began to bellow loudly and seemed much agitated. He examined the hole of water and noticed a rope made of cow's hair in the water with a stone on either end of it. He took it out and the cows ceased to bellow and drank quietly. He brought the rope home and thought no more about it. From that day on, whenever they churned they had a very

large quantity of butter and hardly any buttermilk. After churning a few times and wondering at all the butter, he remembered the rope. He searched for it, found it and threw it into the fire where it burned with loud crackling. He was uneasy and went to the priest and told him his story. The priest told him that he should not have burned the rope, that he was only getting back the butter that had been stolen from him, and as soon as he had got his right, the churn would only yield the normal mount of butter.[49]

It was also commonly believed throughout Ireland that well water could be used for butter stealing purposes, and the cream skimmer was often used to skim the surface of a communal well, or a neighbour's well, on May morning. The first water taken from the well on May morning was regarded as being potent for the purposes of good or evil. Thus, whoever was first to skim the well was believed to have milk and butter luck for the coming year, including also perhaps, that of his neighbour, (or neighbours), if a communal well had been skimmed. People therefore tried to ensure that a member of their own family drew water from their well on May Morning before any other person did so. We hear of wells, particularly those privately owned, being physically guarded from sunset on May Eve to sunrise on May Morning, a custom which was particularly common in the rich dairying areas of the South and East of Ireland. But it was also practised on smaller farms, and the following account from south Co. Clare—apart from detailing the commercial aspect of even small-scale butter-making—shows the context in which belief in butter theft by means of magical interference with the family's water supply could be actualized:

My father was a farmer's son. His father's farm ... carried about twelve cows, and in addition to the farm proper they had a large tract of mountain. One year they utterly failed to make any butter. The cream was skimmed off the pans as was the custom then, and put into the barrel- or hand-churn. They then made an effort to churn it into butter as usual, but without a satisfactory result. The cream, with all the churning, turned into a gaseous frothy mass. Everyone in the house took a hand with the churning, but all their combined efforts effected no change in the cream. Eventually it was given to animals.

Now, at that time and for years after, the custom was that 'partners' filled firkins of butter. That meant that the farmers' wives churned once a week, and each brought her quantity of butter, small or big, to our house for that week. There the whole quantity of butter was put into a tub and mixed, and eventually put into the firkin for conveyance to the market. Small farmers with three or four, or even five cows in bad land, never had more than one firkin for the market. Larger farmers had two, while a farmer able to market three firkins of butter together was very snug indeed. When the firkin or firkins were filled the visitors were entertained to tea, and after a little delay for gossip, then left for home with their empty butter vessels.

Now during this period under review my grandparents had no butter to give, and consequently no butter to get. A small farmer of four cows lived near them. Very frequently he passed over my father's 'street'[50] on the way to the market with *two* firkins of butter on the car, and his wife perched up on the seat as proud as punch. This fact gave rise to thought. One May morning my grandfather hid near a spring well in his land. About four o'clock he saw the woman approaching the well. On peeping out stealthily he found she was [the woman who had] the two firkins for the market out of the small farm. She approached the well and started the incantations. My grandfather didn't let her proceed far when he appeared. There was a scene of course. He got her to admit that she stole, and was then making an effort to steal, his butter. When he threatened exposure and publicity, she vowed to give up the practice on condition that he kept her secret. His butter-making came alright after that. This is a perfectly true fact.[51]

Communal wells presented a particular danger to those who used them, but in some areas the person on whose land the well was situated had the right to take the first water from the well on May morning.[52] Such wells, like those privately owned, might also be guarded as the following account from Co. Waterford shows:

At Mooneire, at the well used by the people of the hamlet of Seskin, the local farmers used to watch together. [A man] told me recently that his father, [and a number of other men] were watching, and getting thirsty agreed to go for a drink. [One of the men] tried to skim the well on the sly, but [another man] caught him at it and knocked him into the well with his ashplant.[53]

In the foregoing account it is, as far as the material to hand shows, atypically, a man who tries to skim the well. Generally speaking, however, the material points to women as the milk thieves, and some families had even a reputation for milk stealing, the power being passed on from mother to daughter one such family in Co. Leitrim was nicknamed 'pull the rope'!). In a narrative from the rich dairying Golden Vale area of Munster about a woman attempting to set charms at a neighbour's well on May morning, the woman's daughter is also said to have been involved in milk-stealing activities, which she engaged in in hare-form:

A daughter of this woman . . . was, on another occasion, the subject of *piseoga*. On May eve a dog chased a hare and the hare ran towards this woman's house, and was jumping through the window, when the dog caught her and bit her. The woman wore the mark of the bite ever after.[54]

218 Milk and milk products

The foregoing is a version of a migratory legend found throughout Ireland which substantiates the belief—articulated already in the Irish context by Giraldus Cambrensis in the twelfth century[55]—that some women were capable of shape-shifting and could transform themselves into a hare in order to suck the milk of their neighbour's cows. The following version is from Co. Clare:

> Some people would milk their neighbour's cows on May morning. One such woman was shot at here when she was milking. She turned herself into a hare. She was fired on as such and hit. She was tracked by blood and found bleeding in bed beside her husband.[56]

In other versions of the legend the 'hare' is forced to regain human female shape when attacked by a jet black hound (i.e. 'without one single white hair'), as in the following example from Co. Longford in the midlands of Ireland:

Milking Hare

> A story is told of a hare that used to be seen milking cows—I don't know was it on May morn or was it a frequent occurrence. She was chased by hounds on many occasions but always got away. A strange travelling woman told that the hare could be caught only by a black greyhound—completely black without even one white hair. Such a dog was found after a long search, the hare was waited for, arrived, and the chase commenced. All the other dogs in the district—of every breed and class—took part, but as the hunt was a long and fast one, all dropped out or were left behind except the black one who was closing in on his quarry. The hare, being pressed, made towards a little house where an old woman . . . lived alone, and was escaping through the window when the black greyhound snapped at her and took a piece of flesh off her hind leg. There was no other exit except the door which was closed. The hunters were in view of the window, and on arrival barred it up, and entered the house to search for the hare. The old woman was spinning in the corner and could give no information. After searching fruitlessly some of the hunters became suspicious, and on examination, found a wound on the woman's leg still bleeding and exactly corresponding with the piece of flesh the dog had torn off. It is not recorded what became of her but a family . . . is still [nick] named *Girreys* [< Ir. *giorria*, 'hare'].[57]

So far, milk stealing rites at boundary water, in pastures, or at household or communal wells, have been considered. But a whole range of milk and butter-stealing rites are also linked to the farmyard and the dwelling-house. Opening all the farmyard doors and gates was a way of letting the farmer's luck escape according to a Co. Carlow correspondent writing in 1947:

Another belief is that the luck is taken by a person coming during the night of May eve and opening all the doors and gates of the farmyard. I am reliably informed by a local that he saw the opening performed once, and could not know what it meant until some older people made him wise.[58]

The charm-setter might also use sympathetic magic to take milk and butter luck. This was done by appropriating dairying objects or utensils belonging to the farm, or by taking cow hairs, mud from the cloven foot, cow droppings, or especially the spancel from the byre. Spancels so stolen might be used for mock or symbolic milking in order to take a neighbour's profit, as the following account from Co. Fermanagh shows:

> There are also stories of milking the rope. An old hag who lived at Coragh was spied upon, and was seen with a rope suspended from a beam at the end of the churn and she milking it. She too was noted for the amount of butter she sold in relation to the number of cows she had.[59]

Appropriating, or obtaining by means of a trick, milk, butter or a milk vessel from a house on May morning could also put the family's butter supply for the coming year in jeopardy, according to popular belief. This was particularly so if a May morning visitor did not help with the churning while it was in progress. Indeed so great was the fear of being blamed if the churning failed that people were very reluctant to visit, or be seen near, their neighbour's house on May Day. It was also strongly believed throughout Ireland that a person could steal a household's butter by taking fire—symbolic of the household's luck, which on May Day centres on milk and dairy produce—out of the house on that day. Thus, even a frequent pipe-smoking caller to the house would be refused a live cinder to 'redden' his tobacco pipe on May Day, and indeed he would be expected not to ask for it. The following legend, while hinting at the carelessness of the housewife who left her churn unattended, tells how a tailor foiled the attempts of a butter thief attempting to steal butter by means of a live coal from the household fire:

> A travelling tailor was one May day engaged in a certain farmhouse. The woman of the house was churning, and once while she left the house a strange woman entered and took the coal from the fire. No sooner did the tailor see this than he picked up a coal and dropped it into a pail of water. It seems that this had the effect of quenching the coal the woman had taken because she presently returned and took another glowing coal. The same operation performed by the tailor resulted in her returning again. Her next attempt also failed, and by this time she understood that some power greater than her own was at work, and so she departed, but through the sharpness of the little tailor, without the butter she intended to bring.[60]

220   *Milk and milk products*

Even the smoke from the kichen fire could represent cream, and the household's butter could be stolen by the charm-setter who recited *im an deataigh sin ar mo chuid bainne-se*, 'the butter of that smoke on my own milk', while reversing into her own house. Consequently people were reluctant to be the first household in their townland to light a fire on May morning for fear of drawing the attention of charmsetters. *Codladh Bealtaine* ('May morning sleep') was an expression applied to late rising on May morning in parts of Connacht, and indeed a Co. Galway correspondent tells us that:

> The women of south Conamara (along the seashore) would not light a fire on May morn until they saw smoke from the houses in Co. Clare, because they were afraid that the Clarewomen would take the butter from them across the sea.[61]

Thus people tended to do outdoor work such as hedging in sight of each other's homes on May Morning in order to keep watch for smoke rising from houses, and this activity is the basis of the remark 'he is May morning hedging', to describe a person who is spying on his neighbours.[62]

Protecting the milk profit

In view of such a strong belief that milk and milk products were seriously at risk during the festival of May, it is to be expected that people would take measures to protect them, and it is quite clear from the source material that a wide variety of protective measures were known and practised during the festival for that very purpose. Some of these derived from official religion, such as having a Mass said in the family home, sprinkling holy water and especially Easter water on the farm, on the farm animals, and on the milking and churning utensils, or hanging religious medals or other sacred artifacts in the cow byre. But it is also clear that these measures were not always considered to be sufficient to counteract the perceived threat to dairy produce at Maytime, and thus recourse was had to other, or additional measures of an a-Christian nature, for protective purposes. These measures had the sanction of tradition, and even though they were recognised as acts of magic, people felt justified in performing them in order to protect milk and butter which were a vital food resource, and perhaps also, a source of income.

Only a small sample of the extensive catalogue of protective measures detailed in the source material can be mentioned here.[63] Included among them are attempts to protect boundary water and well water by the application of iron (often in the form of horseshoe nails) and salt. Twigs of the quicken or rowan tree were placed at the four cardinal points on the farm, or a more formal redefining of the farm boundary might take place.[64] Cows were driven through fire, or a small fire of furze bushes might be lighted in a field near the farmhouse. At the farmyard, all gates and shed doors, especially those of cow-byres, were securely fastened or locked, and milking and churning vessels were protected with twigs of the rowan tree. Care was also taken not to allow any such vessels out of the farmyard on May day, or to give away milk or butter, and also to prevent people from taking live coals or cinders from the household fire on that day. In addition, in the eastern and southern parts of Ireland in particular, a variety of protective domestic verdure symbols including May flowers, the May bough and the May bush were used in farmyard or dwelling house, or at the well, during the festival of May. The May flowers—usually yellow flowers such as marigolds, cowslips, buttercups and furze blossoms—were normally picked on May eve and placed over the kitchen and byre doors, on the thresholds and window sills and perhaps also at the well. The May bough—typical of the province of Munster—was brought into the house and put in a place conspicuous to any would-be milk thief, while the May bush—common in the southeast and north-midlands, and often a whitethorn bush decorated with yellow flowers and eggshells—was placed before the kitchen door or at the farmyard gate. These verdure symbols, placed at liminal points about the farmyard and house, were intended to protect and promote the household's seasonal luck and prosperity concerned with milch cows, pasturage and dairy produce.[65]

If, however, in spite of, or due to a lack of, protective measures, the milk profit was believed to be stolen, further measures could, and should be immediately taken to recover the butter. In the case of repeated failures to make any butter, or a reasonable quantity of it, recourse was often had to the parish priest to bless the dairy and farm utensils, or to say a 'May Mass' in the house or church for the recovery of the milk profit.[66] For an unexpected churning failure on May morning more traditional methods may be used such as putting hairs from a cow's tail on the churn dash, as the following account from Co. Clare indicates:

> [A] family were unable to make a churn until [an old woman came in. She took a *dreas* [a turn at churning] and was unable to make progress. She went out, made a ring [of cow hair and put it on the dash of the churn] and all was well.[67]

The heating of iron—especially ploughing irons—in a strict ritual process intended to reinforce boundaries was very common. This was designed to reveal the identity of the butter thief and thus recover the butter, as outlined in the following legend from Co. Westmeath:

> To get back your butter: bolt the door tight, close the window, put salt in the milk, and put the coulter and plough chains into the red fire, and then start churning. As the irons become red you'll hear someone at the door, but keep churning. Then the person who took your butter will screech to you and beg of you to let her in. But take no notice until

your churning is done and you get the butter. After that this person will have no power over the milk.[68]

Conclusion

The perception that both the knowledge and performance of milk-stealing and protective magic belong essentially to women, is evident from the traditional source material. Men on the other hand are shown to be involved in the theft of milk profit only in exceptional or accidental circumstances, and, as is pointed out in some of the legends quoted above, they are said to take steps to 'return' the 'stolen' milk profit. Indeed a male informant from Co. Cork claims that 'It was never heard of that a man was found trying to steal the produce of a neighbour's herd',[69] and a mid-nineteenth century account from Co. Derry states that a husband died of shame after his wife had been found guilty by the local church minister of milk theft by means of acts of magic.[70] Men's role in relation to the protection of milk and butter luck is concerned mainly with situations in which their physical strength or force is required, such as at the guarding of wells, hunting the milk-stealing hare, or preventing women from gathering the dew. Indeed the use of physical force against women—or at least the threat of it—is a recurring feature in traditions about milk luck protection at Maytime where men are involved.[71] This may reflect prevailing social prejudices which consider women to be more prone to magic than men, with the added effect of shielding men from community criticism.[72] But on balance, the main reason that society has viewed women as milk 'thieves', is undoubtedly linked to the fact that traditionally, it was the woman or women of the household, who had responsibility for milking, butter-making and associated utensils, as well as the care of milch cows especially after calving, and the calves too. In the Irish context there is abundant evidence of the importance of the woman's role in the production of dairy produce, ranging from the ancient law tracts of about the eighth century, to the personal experiences of some Irish farmers' wives in modern times.[73] It was the woman's responsibility to ensure that there was a plentiful supply of milk and dairy produce, both for household use and, also, perhaps for the market. To achieve this it was necessary to have not only the required technical skills in relation to milk and butter production and the care of cows, but also to have knowledge of the performance and effects of milk magic and counter-magic.

Since milk and butter production were the responsibility of the women of the community, and since they no doubt wished to be as successful as possible in this vital undertaking, they were, of course, in competition with each other. However, it was important not to be arrogant about one's butter luck, or to appear too successful. As we have seen in some of the material presented above, if the quantity of butter produced on a farm was considered disproportionate to the number of cows it supported, or the size or quality of the holding, the woman could be suspected of milk-stealing, especially if a neighbouring farmer in ostensibly better circumstances was unable to produce a reasonable quantity of butter. For example, a certain old woman farming bogland in Co. Laois, with a couple of what appeared to be poorly fed cows, who yet had big baskets of butter for sale despite having a family to provide for, was considered to be a butter thief.[74]

But it would also seem that it was not just the marginal location or status of farm holdings which might give rise to suspicions of butter-stealing on the part of some women. While any woman in the neighbourhood was felt to be a potential butter thief at the boundary festival of May—even one's most obliging and friendly next-door neighbour[75]—it is evident that some women in particular were considered a threat at this time. The extra-categorical status of some of the women farmers seems to be a particularly relevant factor. It is evident from the descriptions of milk 'thieves' in the source material—only a small proportion of which is presented here—that many of these women were old, or ugly, and in this connection a Co. Monaghan informant states: 'It is said still of any ugly woman that she was like one 'who would be sweeping the dew'.[76] Others were women who were widowed, unmarried, or independent women farmers living alone, or women of the Protestant faith living amongst old Catholic stock—all thus lacking supportive social ties and, therefore, potentially culpable and easy scapegoats.[77]

Accordingly, there can hardly be any doubt that some farming women became community scapegoats, as a result of attempts by their neighbours to account for variations in the expected return from their milch cows, due perhaps to changes in environmental conditions, fluctuations in milk quality, or mismanagement. From the point of view of those households which had suffered churning failures or a substantial reduction in butter quantity, for whatever reasons, the general belief in the possibility (or indeed, probability) of women stealing butter by magic during the boundary festival of May, enabled them to shift the responsibility for such a situation onto other members of the community—who were almost invariably women. Thus, the ineffectual farmer could escape criticism, and the housewife who had failed in her responsibility to provide a sufficient quantity of butter for the household or market, could de-emphazise her own inefficiency in butter production by placing the blame for her failure on other neighbouring women, some of whom were clearly more competent and prosperous than she, but also, unfortunately, socially disadvantaged.

The festival of May was a time of considerable reorganisation on the farm especially in relation to cattle. The cow herd which had been kept indoors over the winter and spring seasons was now turned out of doors into nearby pastures, or sent to the mountain or moorland *buaile* for the duration of the milking season, if transhumance was still practised. The women's daily routine was also altered to cope with the summer supply of milk, and the substantial and labour intensive butter-making task. As Mayday was considered the beginning of the milking season, it was a time of heightened competition between neighbouring milk producers, when women in general, but some women in particular, were under suspicion both by men and women, as potential milk-profit thieves. It was a time of latent hostility—or even obvious aggression—towards one's neighbours, as people sought to safeguard their livelihood during the dangerous transition period between spring and summer—the boundary festival of May.

Note: IFC—Irish Folklore Collection in the archive of the Department of Irish Folklore, University College Dublin, which includes the questionnaire entitled May Eve and May Day circulated by the Irish Folklore Commission in 1947. Because of the extensiveness of the material it has only been possible in this short paper to give the most pertinent references. Individual and family names as well as some place-names have been omitted from the IFC texts.

NOTES AND REFERENCES

1. A. T. Lucas, *Cattle in Ancient Ireland*, Kilkenny 1989, 3, and note 3, p. 250. For a textual example of the sentence, see M. Freeman (ed.), *Annála Connacht. The Annals of Connacht, (A.D. 1224–1544)*, Dublin 1944, 19, par. 26, l. 4– '. . .₇ *doronsatur sith dar cend a mbo ₇ a muintire* ('. . . and they concluded peace for the sake of their cattle and their people').
2. Lucas 1989, 3.
3. For a critique of these literary texts as ethnological sources, see Lucas 1989, 4–5.
4. In relation to cow-raiding, see Lucas 1989, 125–199. See also O'Rahilly (ed.), *Táin Bó Cúailnge*, Recension 1, Dublin 1976, and works quoted there esp. R. Thurneysen (ed.), *Die irische Helden- und Königsage bis zum siebzehnten Jahrhundert*, II, (Halle 1921), 96–219.
5. R. Thurneysen (ed.), *Cáin Lánamna* (§21), in *Studies in Early Irish Law*, Dublin 1936, 45–6; see also D. Ó Corráin, 'Marriage in Early Ireland', in A. Cosgrove, (ed.), *Marriage in Ireland*, Dublin 1985, 9.
6. In order to work out an equitable division of assets in the case of a 'marriage of common contribution', a marriage in which, apparently, both parties contribute equally to the common pool of marital property the Irish lawyers hit upon the handy notion of a threefold division between land, labour and capital (livestock) (Ó Corráin, *op. cit.*, 8). But in the division of consumables, including dairy products, a further principle was applied by the lawyers—added value; thus, where the production of the product is labour intensive, it is valued in terms of the labour put into its production (rather than the original raw materials), and in the case of dairy products, the woman's share on divorce reflects this (Ó Corráin, *op. cit.*, 8). Therefore, in the division of dairy products in respect of a lawful wife of equal standing and birth as her husband, 'the labour third is divided in two portions and the woman (who of course has run the dairy) takes one; of the remainder (i.e. one sixth of the whole), diminishing fractions go to the spouse who supplied the dairy vessels (a matter of considerable importance, for dairy vessels were expensive artifacts produced by skilled craftsmen), the husband and the spouse who provided the dairy workers.' (Ó Corráin, *op. cit.*, 9; *Cáin Lánamna* (§12) in Thurneysen, *op. cit.*, 31–32). On the other hand, a woman involved in a

'marriage on the man's contribution', in which the bulk of the marriage goods are contributed by her husband, is in a much less favourable position in relation to the division of assets on divorce. Nevertheless, her husband may not alienate cows without her consent, and she receives one sixth of the dairy produce in store (Ó Corráin, *op. cit.*, 9; *Cáin Lánamna* (§27), in Thurneysen, *op. cit.*, 56). In the case of a 'marriage on a woman's contribution', a marriage to which a woman brings the preponderence of the property, the husband on divorce gets one-eighteenth of the dairy produce (Ó Corráin, *op. cit.*, 10; *Cáin Lánamna* (§29), in Thurneysen, *op. cit.*, 57–62).

7 A. T. Lucas, 'Irish Food Before the Potato', *Gwerin* 3, 1960, 19–31, *Passim*.
8 K. Meyer (ed.), *Aisling Meic Conglinne. The Vision of MacConglinne*, London, 1892, 212pp. For the place of milk and milk products in the diet cf. also Lucas, 1960, 19–31; P. Lysaght, 'Continuity and Change in Irish Diet', in A. Fenton, E. Kisbán (eds.), *Food in Change*, Edinburgh 1986, 82–83.
9 Lucas 1960, 20.
10 *Idem*.
11 W. O'Sullivan, *The Economic History of Cork City from the Earliest Times to the Act of Union*, Cork, 1937, 83.
12 *Ibid.*, 104–106.
13 *Ibid.*, 125.
14 *Ibid.*, 154.
15 *Ibid.*, 105.
16 *Ibid.*, 154.
17 *Ibid., op. cit.*, 154–8; casks made of seasoned sycamore and beech also fulfilled the requirements of the Cork Butter Market (O'Sullivan, *op. cit.*, 261).
18 *Ibid.*, Appendix No. 32. a—d, 334–340.
19 J. O'Donovan, *The Economic History of Live Stock in Ireland*, Cork 1940, 303—The heavily-cured butter from the Cork market packed in superior Cork-made casks (O'Sullivan, *op. cit.*, 261-2) rendered it suitable for transportation to distant markets. In the English market the Cork butter was preferred for the winter season because of its keeping qualities, while the butter of Waterford, Limerick and Belfast was—because of its light cure—preferred in summer and autumn.
20 The firkin is a small wooden cask for liquids, butter, fish etc., made by the cooper (*Oxford English Dictionary*).
21 O'Donovan, *op. cit.*, 301.
22 O'Sullivan, *op. cit.*, 147–8, 160: O'Donovan, *op.cit.*, 313.
23 O'Donovan, *op.cit.*, 309; O'Sullivan, *op.cit.*, 256–278, 334–340; 357–361; J. S. Donnelly, Jr. 'Cork Market: Its Role in the Nineteenth Century Irish Butter Trade', *Studia Hibernia*, **11** (1971), 130–163.
24 O'Donovan, *op. cit.*, 317; For the reasons for the decline of the market, see O'Donovan, *op. cit.*, 315–317; L. Kennedy, 'The Decline of the Cork Butter Market: A Comment', *Studia Hibernia*, **16** (1976), 175–177; Donnelly, *op.cit.*, 143ff.
25 O'Donovan, *op. cit.*, 319–20; P. Doyle, L. Smith, *Milk to Market*, Dublin 1989, 160.
26 O'Donovan, *op. cit.*, 320–321.
27 *Ibid.*, 321.
28 Doyle, Smith, *op. cit.*, 114.
29 *EC Dairy Facts and Figures*, Surrey 1991, 30, Table 13, (12 countries).
30 *Ibid.*, 60, Table 31, (1989 figure, 10 countries).
31 *Ibid*, 170, Table 114, (9 countries). In world terms Ireland ranks fourth after Iceland, Finland and Norway; Source AMS. 1991, Tetra Pak.
32 *Ibid.*, 171, Table 114, (9 countries).
33 *Ibid.*, 174, Table 116, (10 countries).

34 *Ibid.*, 175, Table 117, (10 countries).
35 Lucas 1989, esp. 11–14.
36 The belief that old women in the form of hares sucked milk from cows is mentioned by Giraldus Cambrensis towards the end of the twelfth century—see note 55 below. An account from the middle of the sixteenth century links this belief to May Day and mentions other May Day beliefs well known in modern times in Ireland—in this connection, see K. Danaher, *The Year in Ireland*, Cork/Dublin, 1972, 109–110.
37 For the festival of May see Danaher 1972, 86–127; C. Ó Danachair, 'The Quarter Days in Irish Tradition', *Arv* **15**, (1959), 49–50, 53–55; R. Buchanan, 'Calendar Custom', *Ulster Folklife*, 1962, 24–30; E. E. Evans, *Irish Folk Ways*, London 1957, 272–4; P. Lysaght, 'Maytime Verdure Customs and their Distribution in Ireland', *International Folklore Review*, London 1991, 75–82; P. Lysaght, Bealtaine: 'Irish Maytime Customs and the Reaffirmation of Boundaries', in H. E. Davidson (ed.), *Boundaries and Thresholds*, Woodchester, 1992, 28–43; P. Lysaght, *Féile na Bealtaine in Iarthar agus Iar-Dheisceart Thír Chonaill* ('The May Festival in West and Southwest Donegal'), forthcoming in B. Almqvist (ed.), *An Chéad Céad. The First Hundred. Festschrift for Séamus Ó Catháin*, Department of Irish Folklore, Dublin, 27–41. See also H. Glassie, *Passing the Time*, Dublin 1982, 778, note 1, and A. and B. Rees, *Celtic Heritage*, London 1961, Ch. 3.
38 Cf. Lysaght 1991, 76–77, 1992, 31. Although a large body of material was collected as a result the questionnaire about May Eve and May Day in 1947, a few collectors experienced difficulties because of some peoples' reluctance to admit familiarity with the Maytime magical rites for fear of being thought to practise them (IFC 1095: 280, Co. Tipperary).
39 IFC 1097: 185.
40 While strangers might occasionally be blamed for butter theft, the weight of the evidence is that it was envious neighbours who were believed to be involved. A relative is specifically mentioned in IFC 1095:145, Lismore, Co. Waterford, for example. The belief that women were involved in milk profit theft by means of magic acts is also found in other cultures as is evident from the proceedings of witch trials. For the situation in the Nordic countries cf. J. Wall, *Tjuvmjölkande väsen*, 1, Studia Ethnologica Upsaliena III, Ch. 4.
41 IFC 1096: 83 Co. Mayo.
42 IFC 1096: 8. (*Ceaptaí (agus ceaptar i gcónaí) gur droch-fhuadar a bheadh faoin duine a rachfadh ar talamh duine eile Lá Bealtaine i.e. chun piseoga a chur ag obair.*).
43 IFC 1095: 114–115.
44 For the composition and construction of the *buarach*, see Lucas 1989, 44. Lucas has also presented a wide-ranging survey of early and medieval Irish literary texts to show that from earliest times milking was regarded as woman's work, and that the spancel was regarded as her indispensable accessory in that task.
45 IFC 1096: 341. Cf. Lysaght 1993, 32–33.
46 Translation of Irish text from the bilingual collection of legends: S. Ó hEochaidh, M. Ní Néill, S. Ó Catháin, *Síscéalta Ó Thír Chonaill /Fairy Legends from Donegal*, Dublin, 1977, No. 25. This legend expresses the common idea that alien population groups—and in this case with an alien faith—had occult powers. For a further version of this legend from southwest Donegal expressing a similar view of Protestant women, see IFC 1096:407–8.
47 IFC 1097: 81; it would seem as if virtually any object could become a medium for magical milk-stealing on May morning provided it was used in the prescribed manner.
48 The reference to black magic is contained in the threat 'to steep the sheaf' or 'to bury the sheaf' as the magical activity is more usually called. The sheaf represented the intended victim, and pins might be stuck in the joints of the

straw wisps in order to give a painful death. The sheaf was buried (or thrown into water in an inaccessible place such as a boghole), and as it decayed, the person it represented was also believed to fade away. This custom was strong in the eastern midlands of Ireland settled by English colonists, including Co. Westmeath, from where the legend cited comes. For an analysis of this phenomenon, see N. McLaughlin, *Burying the Sheaf. A Form of Murder by Magic*, student essay in the Department of Irish Folklore, 1977.

49 IFC 1095: 35.
50 'Street' is the yard in front of the farmhouse in west-Co. Clare.
51 IFC 1095: 236–7. Cf. Lysaght, 1993, 33–34.
52 IFC 1196: 407, Ardara, Co. Donegal.
53 IFC 1095: 335.
54 IFC 1095: 278 Tipperary.
55 *Giraldi Cambrensis Opera*, Vol. V. ed. James F. Dimock, London 1867, 106; '*Item, vetulas quasdam, tam in Wallia quam Hibernia et Scotia, se in leporinam transmutare formam, ut adulterina sub specie ubera sugendo, lac alienum occultius surripiant, vetus quidem et adhuc frequens querela est.*' For a mid-sixteenth century reference to the belief that old women in hare-form sucked milk from cows on May Day, see the section entitled 'Ireland and the Smaller Ilands in the British Ocean', in W. Camden, *Brittania*, London, 1610, 146: '... If they finde an hare amongst their heards of cattaile on the said May daie; they kill her, for, they suppose she is some old trot, that would filch away their butter ...'; also quoted in Danaher 1972, 109–110. For nineteenth and twentieth century references see T. Crofton Croker, *Researches in the South of Ireland*, London 1824, 94 (Also facsimile edition, Shannon, 1969, 94), and S. Ó Duilearga, *Leabhar Sheáin Í Chonaill*, (Baile Átha Cliath 1948, 1971), 437/*Seán Ó Conaill's Book*, Dublin 1981, 399–400; P. Lysaght, 'A Tradition Bearer in Contemporary Ireland', in L. Röhrich/S. Wienker-Piepho (eds.), *Storytelling in Contemporary Societies*, Tübingen 1990, 208
56 IFC 1095: 199.
57 IFC 1097: 151. For a Nordic perspective on milk-stealing creatures, see J. Wall, *Tjuvmjölkande väsen*, 1, 2, Studia Ethnologica Upsalienia III, 5, 1977, 1978.
58 IFC 1097: 214.
59 IFC 1096: 341, 337; see also a Co. Waterford account of an old woman milking a *buarach* placed across the fire crane into a bucket (IFC 1095: 317).
60 IFC 1095: 227–8. See also Ó Duilearga, *op. cit.*, 1948, 1971, 310–311, 437; 1981, 272, 400
61 IFC 70: 57.
62 IFC 1096: 341.
63 See Danaher 1972, 109–119; Lysaght 1992, 37–42.
64 For a formal redefining of boundaries described in a nineteenth century source, see Ó Danachair 1972, 116–117.
65 Danaher 1972, 88–9; Lysaght 1991, 77–82; 1992, 39–41.
66 For example IFC 1095: 90, Co. Cork.
67 IFC 1095:200.
68 IFC 1097: 66.
69 IFC 1095: 134; Cf. Lysaght 1993, 36–37.
70 Danaher 1972, 115–116.
71 Indicative of this is the following remark allegedly made by a missioner at the Parish mission in Cappawhite, Co. Tipperary in 1936, and reported in IFC 407: 286: Missioner: 'I'm told that the man who fired on the woman who was taking his butter by *piseoga* only wounded her—more's the pity!'. In addition, a Co. Tipperary questionnaire correspondent wrote in 1947, that there had been a law case in Cashel a few years previously brought by a woman who was assaulted by a neighbour at the well on May Morning (IFC 1095: 269). Also from Co.

Tipperary it was reported that a man who encountered a woman in the field on May morning 'taking' milk with a *buarach*, took the *buarach* from her and flogged her with it (IFC 1095-254).
72 According to Marcel Mauss, women, because of society's attitude to them, are everywhere considered more prone to magic than men: M. Mauss, *General Theory of Magic*, (trans. R. Brain), London, 1972, 28. (Engl. edn.).
73 See note 44 above, and Ó Corráin, *op. cit.*, 9, who writes in reference to the value of the law tracts as ethnological evidence: 'The legal tracts incidentally provide first class evidence of the importance of a woman's role (as manager and worker) in the rural economy—in dairying, in the production of woollen and linen garments, in caring for farmyard animals (especially the fattening of stall-fed beasts for the table), and in organising the ploughing and reaping of corn, (and, no doubt, the feeding of the labourers)'. The gender-based bias in favour of women as milkers still persists to some extent in Ireland in farms where the number of cows is small, or where a milking machine has not been installed.
74 IFC 1097: 186, (1947). Families living in marginal or remote locations who were successful butter producers might become suspect, like a family on Inniskeen Island, Co. Fermanagh, about whom the questionnaire correspondent commented: 'I myself remember the last members of this family and they were certainly queer' (IFC 1096:337).
75 IFC: 1096:419-22.
76 IFC 1096: 367, 1947). Cf. in this connection, C. Larner, *Enemies of God. The Witch-hunt in Scotland*, Oxford, 1981, 9: 'The witch is old, ugly, and female in most societies.'
77 R. P. Jenkins, 'Witches and fairies: Supernatural Aggression and Deviance', in P. Narváez (ed.), *The Good People: New Fairylore Essays*, London/New York, 326-7. This is evident from a number of the extracts from the archival material presented in this paper. It is also evident from the sources that some of these women had a general bad reputation and they would be specially watched on occasions like the May festival. (IFC 1097: 199, Co. Laois). A correspondent also states that some of them were blamed without foundation (IFC 1095: 170, Co. Cork)

SELECT BIBLIOGRAPHY

F.A. Aalen, 'A Note on Transhumance in the Wicklow Mountains', *Journal of the Royal Society of Antiquaries of Ireland*, **93** (1963), 189-190.

R.A. Anderson, *With Horace Plunkett in Ireland*, London, 1935.

K. Danaher, *The Year in Ireland*, Cork/Dublin, 1972.

J. S. Donnelly, Jn., 'Cork Market : Its Role in the Nineteenth Century Butter Trade', *Studia Hibernia*, **11** (1971), 130-163.

P. Doyle, L. Smith, *Milk to Market. A History of Dublin Milk Supply*, Dublin 1989, 181pp.

E. E. Evans, 'Bog Butter : Another Explanation', *Ulster Journal of Archaeology*, **10**(1947), 59-62.

E. E. Evans, 'Dairying in Ireland through the Ages', *Journal of the Society of Dairy Technology*, **7**(1954), 179-187.

A. Gailey, 'Cultural Connections and Cheese', *Folklife* **25**(1986-87), 92-99.

J. M. Graham, 'Transhumance in Ireland', *Advancement of Science* **10**(1953), 74-79.

D. Houston, *The Milk Supply of Dublin: Report of the bacteriolical investigation of the City of Dublin milk supply*, Dublin 1918. (Preface by Oliver St. John Gogarty).

L. Kennedy, 'The Decline of the Cork Butter Market: A Comment', *Studia Hibernia* **16**(1976), 175-177.

A. T. Lucas, Irish Food Before The Potato, *Gwerin* **3**, No. 2, 1960, 19–31.
A. T. Lucas, *Cattle in Ancient Ireland*, Kilkenny, 1989, 315pp.
P. Lysaght, 'Continuity and Change in Irish Diet', in A. Fenton, E. Kisbán, *Food in Change*, Edinburgh, 1986, 80–89.
P. Lysaght, 'Innovations in Food—The Case of Tea in Ireland', *Ulster Folklife* **23** (44–71).
K. Meyer, *Aislinge Meic Conglinne*. The Vision of MacConglinne, London, 1892.
C. Ó Danachair/N. Ó Dubhthaigh, 'Summer Pasture in Ireland', *Folklife* **22** (1983–84), 36–54.
C. Ó Danachair, 'Traces of the Buaile on the Galtee Mountains', *Journal of the Royal Society of Antiquaries of Ireland*, **75**(1954), 248–252.
J. O' Donovan, *The Economic History of Live Stock in Ireland*, Cork 1940, esp. Ch. XVI (The Dairying Industry), 301–347.
S. Ó hEochaidh, Buailteachas i dTír Chonaill', ('Transhumance in Donegal'), *Béaloideas* **13**(1943), 130–158.
P. Ó Moghráin, 'Some Mayo Traditions of Buaile', *Béaloideas* **13**(1943), 161–172;
P. Ó Moghráin, 'More Notes on the Buaile', *Béaloideas* **14**(1944), 45–52.
M. Ó Sé, 'Old Irish Cheeses and Other Milk Products', *Journal of the Cork Historical and Archaeological Society*, **53**(1948), 82–87.
W. O'Sullivan, *The Economic History of Cork City from the Earliest Times to the Act of Union*, Cork, 1937. (Extensive bibliography of primary and secondary sources).
J. K. Thompson, M. A. Collins, D. E. Johnston, 'From the Caucasus to Ulster—The Co. Fermanagh Buttermilk Plant', *Ulster Folklife*, **34**(1988), 54–59.